Unconventional Photoactive Solids

Institute for Amorphous Studies Series

Series editors

David Adler†
Massachusetts Institute of Technology
Cambridge, Massachusetts

and

Brian B. Schwartz
Institute for Amorphous Studies
Bloomfield Hills, Michigan
and Brooklyn College of the City University of New York
Brooklyn, New York

DISORDER AND ORDER IN THE SOLID STATE
Concepts and Devices
Edited by Roger W. Pryor, Brian B. Schwartz, and
Stanford R. Ovshinsky

DISORDERED SEMICONDUCTORS
Edited by Marc A. Kastner, Gordon A. Thomas, and
Stanford R. Ovshinsky

LOCALIZATION AND METAL–INSULATOR TRANSITIONS
Edited by Hellmut Fritzsche and David Adler

PHYSICAL PROPERTIES OF AMORPHOUS MATERIALS
Edited by David Adler, Brian B. Schwartz, and Martin C. Steele

PHYSICS OF DISORDERED MATERIALS
Edited by David Adler, Hellmut Fritzsche, and Stanford R. Ovshinsky

TETRAHEDRALLY-BONDED AMORPHOUS SEMICONDUCTORS
Edited by David Adler and Hellmut Fritzsche

UNCONVENTIONAL PHOTOACTIVE SOLIDS
Edited by Harvey Scher

A Continuation Order Plan is available for this series. A continuation order will bring delivery of each new volume immediately upon publication. Volumes are billed only upon actual shipment. For further information please contact the publisher.

Unconventional Photoactive Solids

Edited by
Harvey Scher

BP Research International
Research Center Warrensville
Cleveland, Ohio

Plenum Press • New York and London

Library of Congress Cataloging in Publication Data

International Conference on Unconventional Photoactive Solids (2nd: 1985: Cleveland, Ohio)
  Unconventional photoactive solids: proceedings of the Second International Conference on Unconventional Photoactive Solids, held September 9–12, 1985, in Cleveland, Ohio / edited by Harvey Scher.
    p.   cm. –(Institute for Amorphous Studies series)
  Bibliography: p.
  Includes index.
  ISBN-13:978-1-4612-8052-1   e-ISBN-13:978-1-4613-0727-3
  DOI: 10.1007/978-1-4613-0727-3

  1. Solid state physics–Congresses. 2. Solid state chemistry–Congresses. I. Scher, Harvey. II. Title. III. Title: Photoactive solids. IV. Series.
  QC176.A1I556   1985   88-23168
  530.4′1–dc19   CIP

Proceedings of the Second International Conference on
Unconventional Photoactive Solids, held September 9–12, 1985,
in Cleveland, Ohio

© 1988 Plenum Press, New York
Softcover reprint of the hardcover 1st edition 1988

A Division of Plenum Publishing Corporation
233 Spring Street, New York, N.Y. 10013

INTRODUCTION

The papers in this volume were presented at the Second International Conference on Unconventional Photoactive Solids held at the R&D Center of BP America September 9-12, 1985. It was part of an on-going series of conferences with the main aim of stressing the interrelationship of solid state physics and solid state chemistry.

The choice of topics covered a broad range of light-induced solid state phenomena with particular emphasis on novel materials and/or novel phenomena. Organic solids, in particular, were emphasized as they are a natural meeting point of solid state physics and chemistry. A general trend in solid state physics is to more complex materials (e.g. nonequilibrium glassy films, complicated unit cells, extended molecular building blocks, etc.). This trend necessitates the closer interaction between physicists and chemists.

This conference reflects this trend quite dramatically. It is a new grouping together of a mix of materials, people and experimental approaches.

A typical new theme pulling together this new mix can be seen in Part I. Fractals in Disordered Media. A variety of disordered media give rise to unusual temporal patterns of diffusion and reactions. The more familiar spatial patterns of self-similarity are discussed in the first article by M.F. Barnsley.

Another topic pulling together a diverse group of scientists is covered in Part II. Spectral Hole-Burning. A number of leaders in this field are represented in these three papers.

The key topics of disorder, transport, microscopic interactions and their interrelationship are repeated in Parts III - VII. The rich variety of different materials, techniques, and processes are clearly exhibited in these presentations. Applications resulting from this variety are shown in Part VIII.

It is believed that this series of conferences, with this new grouping, can be expected to make a unique impact on the research activities of the participants and hence serve to create an essentially new field. The present volume is offered in the hope of enhancing this aim.

Harvey Scher

CONTENTS

SECTION IV: ELECTRON TRANSFER IN DISORDERED MEDIA

SECTION V: ENERGY TRANSFER IN ORGANIC SOLIDS

SECTION VI: SILICA GLASSES

SECTION VII: CARRIER DYNAMICS IN AMORPHOUS SEMICONDUCTORS

SECTION VIII: DEVICES AND APPLICATIONS

FRACTAL IMAGE ENCODING

M.F. Barnsley, J.H. Elton, A.D. Sloan, and H. Strickland

Georgia Institute of Technology
School of Mathematics
Atlanta, GA

1. INTRODUCTION

Increasingly, the output of physical and numerical experiments is presented as two dimensional images, instead of as tables and graphs of observed real variables. Instances include pictures of diffusion limited aggregates, fractal fingering boundaries between fluids, and images of turbulent flows. One approach to understanding the intricate geometries of such experimental data is traditional: try to isolate physically meaningful real parameters, such as fractal dimension (Mandelbrot, 1982), by passing straight lines through data points derived from the pictures. Another approach is to attempt to approximate the image with a geometrical entity whose structure is, despite appearances, quite simple. Straight lines, squares, circles and other fundamentally Euclidean models will not suffice for this purpose when the data possesses structure which cannot be fully resolved or simplified by rescaling (i.e., magnification). The first purpose of this paper is to describe in simple terms how fractal geometries may be succinctly understood in terms of iterated function systems and to explain how one can go about finding a fractal model to fit given two dimensional data. The second purpose is to relate attractors for iterated function systems to attractors for cellular automata (Wolfram, 1983). The latter is motivated by the following question: Once one has found an iterated function system encoding of an image, can one deduce a set of cellular automata rules which would allow one to construct the image of a lattice? The answer is "Yes", which delights us for this reason. For some time, via the Collage Theorem (Barnsley, et al., 1984), one has known how to make iterated function system models for ferns and leaves; now one may also model the manner in which these forms both grow and stop growing.

1

One of the motivations for the present work is the idea that pictures
are data; pictures of clouds, dendritic crystal growths, profiles of mountains and close-ups of foliage. Rather than viewing such data as a random
assemblage of unrelated systems, one aims to find a deterministic model for
the data as a whole.

In §2 we describe iterated function systems and the Collage Theorem
in terms of an inverse problem: how can one find a set of rules to make a
given image? In §3, we give the relationship of §2 to cellular automata by
presenting a result on the existence of unique attractors for non-local
cellular automata consisting of 'pixel reduction' maps.

2. THEORY

The following is a simplified discussion of material in (Barnsley and
Demko, 1985; Barnsley, et al. 1984; Diaconis and Shashahani, 1984; Hutchinson,
1981).

To approach the inverse problem we begin with the chaos game. Let K
be the unit square $[0,1] \times [0,1] \subset \mathbb{R}^2$, or the screen of the graphics monitor
of a microcomputer, and let there be two continuous mappings, W_H and W_T,
the 'Heads' and 'Tails' functions, both taking K into itself.

A starting point $x_0 \in K$ is chosen; a coin is tossed; and then $x_1 =$
either $W_H(x_0)$ or $W_T(x_0)$ according as the result is heads or tails. The
process is repeated with x_1 the new starting point; and thus a sequence
$\{x_n\}$ is generated. If the maps are contractions, and if all points are
plotted after say fifty iterations, they will distribute themselves approximately upon an interesting, typically fractal set which we call the
attractor A of the iterated function system $\{K, W_H, W_T\}$. With probability
one A is the set of all accumulation points of $\{x_n\}$. It is independent of
the sequence of coin tosses, and even of a possible bias on the coin.

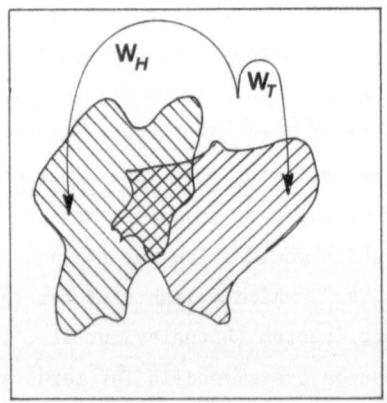

As an example let K be a large piece of the Cartesian plane which includes the interval [0,1] on the x-axis. Let

$$W_H(x,y) = (x/3, y/3) \quad \text{and} \quad W_T(x,y) = (x/3 + 2/3, y/3).$$

Then A is the classical Cantor set in [0,1] obtained by the successive removal of open middle-third subintervals: each successive iterate, whether heads or tails, is closer to [0,1]; once in [0,1] one remains there because $W_{H,T}$: [0,1] ↺ ; and if one is any middle third subinterval, either map takes one out of it, never to return.

More generally, let K be a compact metric space with metric $d(\cdot,\cdot)$, and let W_n: K ↺ be a contraction mapping, with for some $0 \le s_n < 1$

$$d(W_n(x), W_n(y)) \le s_n d(x,y)$$

for each $n \in \{1,2,\ldots,N\}$. Let $p = (p_1, p_2, \ldots, p_N)$ denote strictly positive probabilities with $\Sigma p_n = 1$.

THEOREM (Dubins and Freedman, 1966). There is a unique set $A \subset K$ such that

$$A = \bigcup_{n=1}^{N} W_n(A).$$

Moreover, with probability one, A can be found by the coin toss algorithm.

Notice how A is a union of continuously altered shrunken copies of itself, which shows why A is fractal in spirit. By the last statement in the Theorem we mean this. Start with $x_0 \in K$ and choose successively $x_n \in \{W_1(x_{n-1}), W_2(x_{n-1}), \ldots, W_N(x_{n-1})\}$ with probability p_k attached to $x_n = W_k(x_{n-1})$. Then with probability one $A = \{\text{Lim } x_n\}$.

More complete analysis of the chaos game involves considering the Markov process

$$P(x,B) = \sum_{n=1}^{N} p_n \delta_{W_n(x)}(B),$$

where P(x,B) is the probability of transfer from $x \in K$ to the Borel subset B of K, and $\delta_y(B)$ equals one if $y \in B$ and zero otherwise. There is a unique stationary probability measure μ for the process, and the support of μ in A. The law of large numbers allows us to generate μ by the coin toss algorithm. For full discussion see (Barnsley and Demko, 1985; Dubins and Freedman, 1966; Hutchinson, 1981; Diaconis and Shashahani, 1984).

The above theorem leads one at once to wonder what different sets A one can obtain, say using three affine contractions in the plane; and it is the work of half an hour to write a microcomputer routine to generate some pictures (next page). The results are diverse, fascinating and

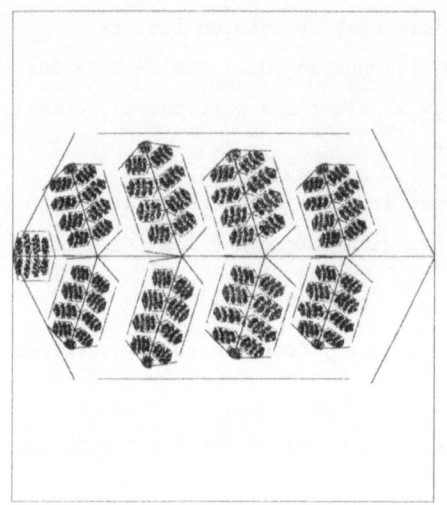

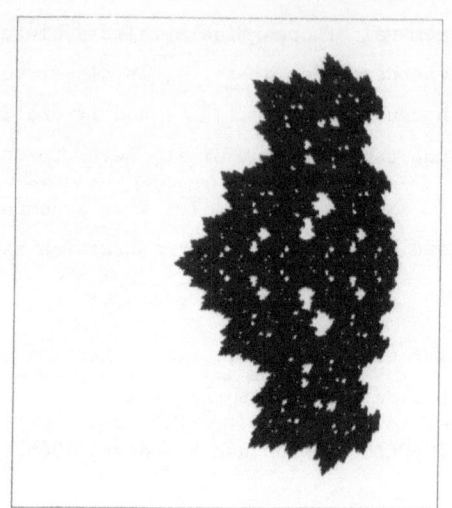

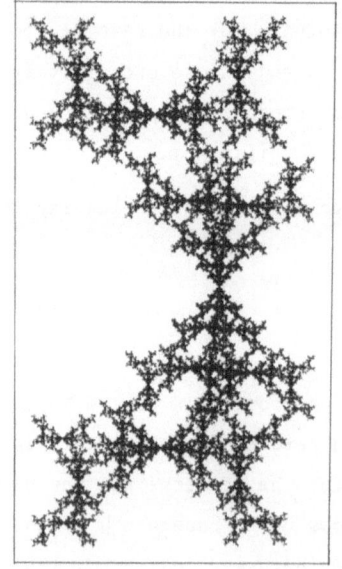

although one may remind one of a cabbage and another of sealing wax, the vast majority appear unfamiliar -- they are of sets never seen in the physical observable universe. We make this last remark because it has application to data compression and objective recognition. It appears that those parts of the appropriate parameter space which correspond approximately to parts of everyday visual images, while locally stable (nearby points give similar pictures), are few and far between. Thus, to recognize a tree one needs to know, only roughly, where one is in parameter space.

We are now ready for the inverse problem. Given a bounded closed set $L \subset \mathbb{R}^2$, representing a picture say, find a collection of contraction mappings $\{W_1, W_2, \ldots, W_N\}$ whose attractor is L. Clearly the task is easy if L is special in the sense that we can immediately spot how

$$L = \bigcup_{n=1}^{N} W_n(L). \qquad (*)$$

For example, let L be the Cantor fan sketched below. Clearly, this is the union of three smaller copies of itself. If we place the origin at the main vertex and the coordinates of some other points are as shown, then (*) is true with N = 3 and

$$W_1\binom{x}{y} = \binom{2/3 \quad 0}{0 \quad 2/3}\binom{x}{y} \qquad W_2\binom{x}{y} = \binom{1/3 \quad 0}{0 \quad 1/3}\binom{x}{y} + \binom{a}{b}$$

$$W_3\binom{x}{y} = \binom{1/3 \quad 0}{0 \quad 1/3}\binom{x}{y} + \binom{c}{d} .$$

Our problem is not with such apparently specialized sets, but rather with the approximation of any set by the attractor of a simple iterated function system.

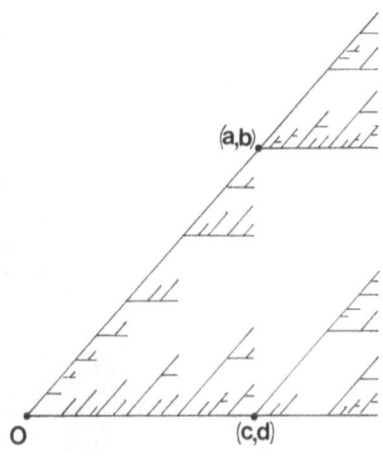

COLLAGE THEOREM (Barnsley et al., 1984). Let the contraction mappings $\{W_n : K \circlearrowleft : n = 1,2,\ldots,N\}$ be chosen such that

$$h(L, \bigcup_{n=1}^{N} W_n(L)) < \varepsilon$$

for some $\varepsilon > 0$. Then

$$h(L,A) < \varepsilon/(1-s)$$

where A is the attractor of the iterated function system.

Here $h(\cdot,\cdot)$ is the Hausdorff metric. It measures the distance between the subsets B,C of K according to

$$h(B,C) = \text{Max}\{\underset{x\in B}{\text{Max}} \underset{y\in C}{\text{Min}} \; d(x,y), \; \underset{y\in B}{\text{Max}} \underset{x\in C}{\text{Min}} \; d(x,y)\}.$$

Also, $0 \le s < 1$ is such that

$$d(W_n(x),W_n(y)) \le s \cdot d(x,y) \quad \text{for all } x,y \in K.$$

This theorem tells us that we need only make an approximate covering or lazy tiling of L by continuously distorted smaller copies of itself, in order to find a suitable set of maps. It also quantifies the continuous dependence of the attractor A on the maps $\{W_n\}$.

In the following example a leaf was approximately covered by four affine images of itself. In complex notation, the maps are

$$W_n(z) = s_n z + (1-s_n)a_n$$

$$
\begin{aligned}
s_1 &= 0.6 & a_1 &= 0.45 + (0.9)i \\
s_2 &= 0.6 & a_2 &= 0.45 + (0.3)i \\
s_3 &= 0.4 - (0.3)i & a_3 &= 0.6 \; + (0.3)i \\
s_4 &= 0.4 + (0.3)i & a_4 &= 0.3 \; + (0.3)i
\end{aligned}
$$

A sketch of the images of L under the maps, and an approximate rendering of the corresponding attractor, obtained stochastically with equal probabilities on the maps, are given next.

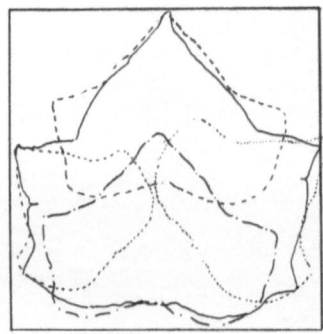 

As another example the reader may like to find the five affine transformations which provide a collage of the following treelike image (below top). Nearby transformations also provide recognizable vegetation images. Short paths in parameter space encode the tree swaying on all scales. The mountains image (below bottom) was obtained by applying the Collage Theorem to several different images.

We remark on the fundamentally parallel nature of the algorithms for regenerating the attractor of an iterated function system from its maps; this means special purpose signal processing hardware can be designed to generate pictures at speeds of the order of one hundred images per second.

The pictures shown here were obtained on an IBM PC with graphics monitor. We have also generated much higher resolution pictures, with delicate shading determined by the invariant measure, using professional computer graphics equipment, see for example (Demko et al., 1985).

We conclude this section by mentioning an area of interesting generalizations. Barnsley et al. (1985) show that unique attractors can be obtained even when not all of the maps are contractions - some can expand - and the probabilities depend upon position. An example of such an attractor is suggested by the following figure which was generated using random iteration of three affine maps, with one an expansion.

3. CELLULAR AUTOMATA

Let L denote a finite lattice and let ℓ, $m \in L$ denote elements of the lattice; as sketched below for example.

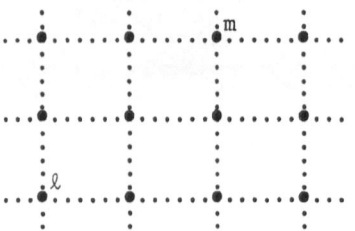

Let $d(\ell,m)$ denote the lattice distance from ℓ to m. Then L with $d(\cdot,\cdot)$ is a compact metric space. Hence if $\{w_i: i = 1,2,...,N\}$ is any set of contractions $w_i: L \to L$ with $d(w_i(\ell),w_i(m)) < d(\ell,m)$ for all $\ell \neq m$ in L, there exists a unique attractor A, a subset of L, such that

$$A = \cup_i w_i(A)$$

Starting from any subset A_0 of L, we have the non-local cellular automata

$$A_n = \cup_i w_i(A_{n-1}) \quad n = 1,2,3,...$$

with the property that $A_n = A$ for all n sufficiently large.

Any set of contraction mappings in $\mathbb{R}^2$, all of contractivity strictly less than 1/2, induces an associated set of contractions on any regular square lattice embedded in $\mathbb{R}^2$. Hence any iterated function system of strict contractions in $\mathbb{R}^2$ may be converted into a cellular automata acting on the screen buffer of a graphics computer.

The following sequence of images shows the development of the Black Spleenwort fern attractor (Demko et al., 1985), starting from a single block of turned on (black) pixels in the upper left corner. The four affine transformations involved may be deduced from the four images of the original block, in the second picture of the series.

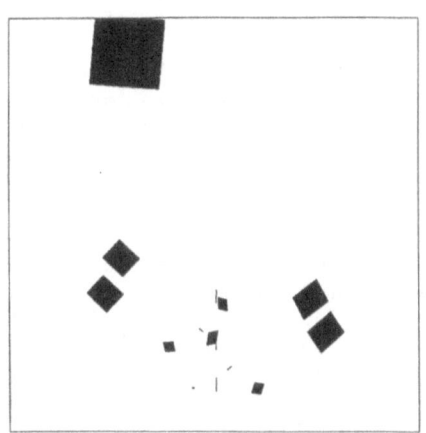

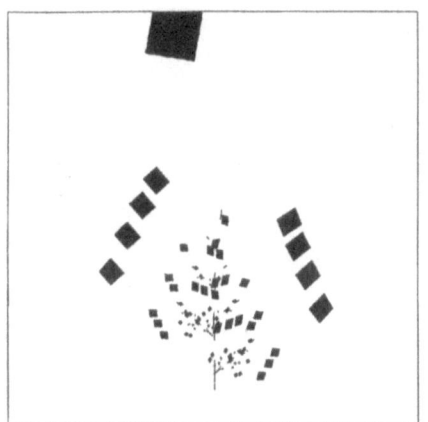

REFERENCES

Barnsley, M. F. and Demko, S. G., 1985, Iterated function systems and the global construction of fractals, Proc. Roy. Soc. Lond., A399, 243-275.

Barnsley, M. F., Demko, S. G., Elton, J. H., and Geronimo, J. S., 1985, On the attractiveness of invariant measures for iterated function systems, preprint.

Barnsley, M. F., Ervin, V., Hardin, D., and Lancaster, J., 1984, Solution of an inverse problem for fractals and other sets, to appear P.N.A.S.

Dubins, L. E. and Freedman, D. A., 1966, Invariant probabilities for certain Markov processes, Ann. Math. Stat. 32, 837-848.

Demko, S., Hodges, L., and Naylor, 1985, Construction of fractal objects with iterated function systems, Computer Graphics 19, 271-278.

Diaconis, P. and Shashahani, M., 1984, Products of random matrices and computer image generation, Stanford U. preprint.

Hutchinson, J., 1981, Fractals and self-similarity, Indiana U. Journal of Math. 30, 713-747.

Mandelbrot, B., 1982, "The Fractal Geometry of Nature," W. H. Freeman, San Francisco.

Wolfram, S., 1983, Statistical mechanics of cellular automata, Reviews of Modern Physics, 55, 601-644.

TRANSPORT AND REACTIONS IN DISORDERED MEDIA

Raoul Kopelman

Department of Chemistry
The University of Michigan
Ann Arbor, Michigan 48109

ABSTRACT

We give an overview of the fractal-like kinetics in geometrically and energetically disordered materials. This includes exciton kinetics on percolation clusters (isotopic mixed naphthalene crystals), grain boundaries of vapor deposited naphthalene films, naphthalene aggregates embedded into "plexiglass" and into pores of vycor glass, polymeric membranes (acetate, nylon) and filter papers (cellulose, glass) with pore sizes from 0.003 to 1.2 microns. The fractal and spectral dimensions derived experimentally for the percolating clusters are in excellent agreement with theory and simulations. The temperature studies separate the effects of energetic and geometric disorder in the naphthalene-impregnated porous materials. Molecular photochemistry in liquid solutions inside pores also shows a fractal-like heterogeneous reaction kinetics (the first such evidence involving a <u>molecular</u> reaction). Supercomputer simulations of reactions demonstrate the subordination law of geometric and energetic disorder. Steady-state supercomputer simulations show an unexpected segregation of reactants on fractal and one-dimensional patches of reactive (catalytic) surface sites. The implication for heterogeneous catalysis, energy upconversion and the characterization of photoactive disordered materials are discussed.

INTRODUCTION

Transport and reactions in fractal-like media have been of much recent interest.[1-13] Exciton transport on <u>percolation clusters</u> was studied experimentally a decade ago[14,15] and has been analyzed and simulated in terms of random walks on percolation clusters[16-18] even before the popular de-Gennes ant first started its drunken walk in a labyrinth[19], and before fractals were introduced into percolation.[20] Our first method of studying mobile exciton transport was based on exciton-trap "reactions". The <u>rate coefficient</u> $\kappa(t)$ of this reaction was related to the random walk efficiency: $\kappa(t) \propto dS/dt$ where S is the range of the random walker (mean number of distinct sites visited)[18,21]. Evesque and Duran[2] used the same approach for the reaction of mobile excitons with trapped excitons (heterofusion) while Klymko and Kopelman[1,22,23] removed the traps altogether and monitored the annihilation (homofusion) reaction between mobile excitons. Both steady-state and relaxation measurements gave anomalous results[23] (e.g. reaction orders of 20 rather than 2); The classical ("euclidean") picture had to be abandoned in favor of a fractal model. This model of geometric disorder appears to have general applicability to disordered media. It is not only complementary to the temporal-disorder (or energetic disorder) model of Scher, Lax, Montroll and Shlesinger[24-28], but the two models can be combined as suggested theoretically by Klafter <u>et al.</u>[8] and via simulations by Anacker <u>et al.</u>[29-31].

We have studied triplet exciton annihilation on naphthalene aggregates which are embedded inside pores[32] (of various membranes or vycor glass), cavities[33] (in polymeric glasses), grain boundaries[34] and on random clusters in random alloys.[1,12,35] The common denominator of all these systems is the non-classical or fractal-like annihilation kinetics. The annihilation rate coefficient κ is not a constant but rather described by the time-dependent form: $\kappa \propto t^{-h}$, where h is a heterogeneity exponent ($0<h<1$) that vanishes ($h=0$) only for homogeneous samples, giving back the classical form: $\kappa=$const. We also measure the dependence of h on temperature. The effective spectral dimension of the fractal-like medium is given by: $d_s=2(1-h)$.

We introduce here on three additional points of interest: 1) A clearcut case in which experiment, simulation and theory agree within three significant figures, giving a convincing demonstration of the usefulness of the "fractal" approach; 2) an example of a photoactivated heterogeneous reaction that proves the fractal-like nature of molecular reactions in heterogeneous media and environments; 3) simulations that demonstrate how binary chemical reactions (including electron-hole recombination) in randomly disordered media may lead to totally unexpected microscopic and macroscopic phenomena.

METHODS

The fusion rate coefficient κ is derived from time-resolved fluorescence (F) and phosphorescence (P) decays[1,36]: $\kappa \propto F/P^m$ but may also be extracted from either F or P alone as functions of time.[12,36] For details see refs. 32-34. Table 1 gives the heterogeneity exponent h and the effective spectral dimension d for various samples and temperatures. We present the data for an assumption of homofusion (m=2) together with the model of heterofusion (m=1), where $k \propto F/P$, (even though this is probably the minor kinetic channel).

RESULTS AND DISCUSSION

We see (Table 1) that all samples exhibit a fractal-like (heterogeneous) behavior (h>0) at low temperatures. There is a reduction in h at higher temperatures. This is consistent with the subordination theorem[8]: $d_s{}' = \beta d_s$ where β is the parameter characterizing the anomalous hopping time distribution (e.g. for continuous time random walk). We also see that some samples have h=0 for higher temperatures. This implies that the "fractal-like" effects are entirely due to energy disorder, so that at higher temperatures the energy disorder parameter W is small compared to T. Obviously our values provide only <u>lower limits</u> for the "real" (geometric) d_s. We note that for vycor

porous glass (at low temperature) $d_s' = 1.1$, which is consistent with the literature range

of values[37] for $d_f \leq 2$ (remembering that for fractals $d_s \leq d_f \leq d$ and where d is the

embedding euclidean dimension). We emphasize that in all these samples, and

especially in the vapor-deposited films, most of the bulk might be quite crystalline (i.e.

euclidean), but our slow experimental time scales assure us that <u>only</u> the sluggish

kinetics in the fractal-like regions contribute to our observations.

For the larger pore membranes, at higher temperatures, we observe the same

behavior as in deposited films, i.e. no delayed fluorescence on the millisecond time-

scale. In short, this approach might characterize crystalline domain sizes. In addition,

we have found excellent correlation between the fractal-like kinetics and a number of

spectroscopic features: 1) The spectral bandwidths (W); 2) Observation of supertrap

(betamethylnaphthalene) emission; 3) Typical photophysical product (excimer) emission;

4) Typical photochemical product (radical) emission[32-34].

A recent analysis[38] of the annihilation kinetics of triplet excitons in percolating

clusters of naphthalene-h_8 in naphthalene-d_8 single crystals gives a refined value of

$h=0.36 \pm 0.03$ and $d_s=1.28 \pm 0.06$. An analysis of the older exciton trapping data of

Ahlgren[10] provided [39] critical percolation exponents ($\beta=0.13 \pm 0.05$ and $\gamma=2.1 \pm 0.2$) from

which the fractal dimension d_f can be derived: $d_s=d(\gamma+\beta)/(\gamma+2\beta)$. We note that the

triplet exciton transport in naphthalene crystals is confined to the <u>ab</u> plane,[39] giving a

euclidean dimension d=2. A comparison of theory, simulation and experiment is given

in Table 2. Note that no free parameters were used to obtain d_f and d_s in Table 2.

The striking agreement of both the fractal and the spectral experimental dimensions

with theory and/or simulation (Table 2) is strong support for the fractal-like kinetics as

opposed to classical kinetics. The naphthalene crystal has been called "the H atom of

molecular crystals". It appears that the isotopic mixed napthalene crystal (at the critical

concentration) is "the H atom of fractal materials". It is the only system that is fractal

Table 1.
Heterogeneity exponents and spectral dimensions.

Sample	Pore Size (μ)	T °K	Exponent (h)		Spectral dim. (d_s)
			m=1	m=2	
Acetate (GA8)	0.2	4	0.38	0.16	1.2 - 1.7
		80		0	3
Acetate (GA1)	0.5	4	0.59	0.47	0.8 - 1.1
Acetate (GA3)	1.2	4	0.57	0.44	0.9 - 1.1
Nylon (B214)	0.2	4	0.42	0.21	1.2 - 1.6
Porous Glass (Vycor)	0.003	6		0.44	1.1
		58		0.5	
Plexiglass (PMMA)		77		0.1-0.3	1.3 - 1.8
Film (Grain Boundaries)		6	0.54	0.45	1.1
Glass filter-paper	0.6	6	0.68	0.42	0.6 - 1.2
		80	0.43	0.20	1.2 - 1.6
Cellulose	0.6	6	0.53	0.33	0.8 - 1.3
		80	0.06		1.9 - 3
Isotopic Mixed Crystals (C_c)		2		0.36±0.03	1.28±0.06

from atomic length scales (10^{-10}m) to finger-size lengths scales (10^{-2}m).

We now describe briefly the first measured fractal-like molecular kinetics. It is a heterogeneous photochemical reaction where two excited molecules in solution dimerize briefly:

$$A^{\dagger} + A^{\dagger} \rightarrow A^{\dagger\dagger}A \rightarrow A + A + h\nu \quad .$$

The rate limiting (by far) reaction is the first (binary) step in which two photoexcited molecules collide, dimerize (excimerize?) briefly, only to end up quickly as ground state monomers plus a photon (uv to blue). We note that the original photoexcitation uses green light, so that this is also a chemical upconversion process. So far this reaction appears to be an ordinary case of homogeneous solution photochemistry. The crucial point is that the solution itself is confined to micro-pores of polymeric membranes (see Table 1). Indeed, experimentally one gets a fractal-like result of h>0 (0.3 being a typical value) rather than the classical answer (h=0). Obviously some of the solvated molecules do not diffuse freely (no classical diffusion constant) but percolate inside the pores (or move along the pore surfaces) with reduced visitation efficiency, giving $\kappa \propto t^{-h}$. We also observe a greatly increased delayed-fluorescence lifetime, compared to the lifetime in a homogeneous solution (for the same temperature, concentration and excitation conditions). Thus, the conditions that led to fractal-like exciton kinetics also cause fractal-like molecular kinetics. Both reactions are monitored in the same way (phosphorescence for the reactant concentration and delayed-fluorescence for the product concentration) and inside the same pore networks.

Table 2.
Experimental vs. Theoretical Dimensions

d=2	Percolation Theory	Percolation Simulation	Experiment	Classical Theory
d_f	1.89583...[*]	1.895±0.002[+]	1.89±0.03	2
d'_s	1.26±0.02[**]	1.25±0.02[++]	1.28±0.06	2

[*]   Stauffer[40] (den-Nuis conjecture).

[**] Mean of Alexander-Orbach[4] and Aharony-Stauffer[3] conjectures, corrected for finite clusters[9].

[+]   Newhouse, Hoshen and Kopelman[41].

[++] Argyrakis and Kopelman[9].

Finally we describe a simulation study on the <u>steady-state</u> reaction A+B → 0 or A+B → AB†, which symbolizes, in principle: 1) electron-hole recombinations, 2) simple bimolecular reactions, 3) matter-antimatter annihilations, 4) heteronuclear fusion, and 5) exciton heterofusion (only in rare cases). In the supercomputer (Cyber 205) simulation. A and B particles parachute onto the sample, <u>at random</u>, with equal rates. The parachuted A or B particle starts an ordinary <u>random walk</u>. Whenever A and B meet, they react (but A + A or B + B do not react). There are no particles at zero time. After a long enough time, a steady state is established. We find that, on a cubic lattice, a steady-state is established fairly quickly, at which time A and B are well-mixed. However, the results of a simulation on a <u>Sierpinski gasket</u> differs in at least two respects: 1) it takes about 3 orders of magnitude longer in time to establish a steady-state (or quasi-steady-state). 2) This quasi-steady state is <u>quasi-segregated</u>. It appears that a mysterious Maxwell-demon is in operation on fractal surfaces. We note that on one-dimensional samples the mysterious segregation effect is also observed, but no quasi steady-state was established (within 3×10^7 time steps).

SUMMARY

We have established a correlation between fractal-like exciton annihilation kinetics and the geometrical constrains and/or energetic disorder of a number of samples containing pure naphthalene. We have shown a behavior analogous to that of isotopic mixed naphthalene crystals (percolation clusters), in contrast to the behavior of perfect, pure naphthalene crystals. Furthermore, for isotopic alloys at the critical percolation concentration, we have demonstrated a genuine fractal behavior, with excellent values for the fractal and the spectral dimensions. For solution reactions inside pores we have given the first demonstration of <u>fractal molecular reaction rates</u>. Our supercomputer simulations show an unexpected trend, namely, the <u>steady-state segregation</u> of reactants on a catalytic substrate. We have thus a new approach to the characterization of transport and reactions for a wide spectrum of unconventional photoactive solids, leading to a new method for the characterization of disordered media.

ACKNOWLEDGMENT

The molecular heterokinetics experiments were carried out by J. Prasad and the supercomputer steady-state simulations by L. Anacker, supported by NSF Grant DMR 8303919 and NIH Grant 2R01 NS08155-16.

REFERENCES

1.   P.W. Klymko and R. Kopelman, J. Phys. Chem. 87, 4565 (1983).

2.   P. Evesque and J. Duran, J. Chem. Phys. 80, 3016 (1984).

3.   A. Aharony and D. Stauffer, Phys. Rev. Lett. 52, 2368 (1984).

4.   S. Alexander and R. Orbach, J. Phys. (Paris) Lett. 43, L625 (1982).

5.   R. Rammal and G. Toulouse, J. Phys. (Paris) Lett. 44, L13 (1983).

6.   P.G. de Gennes, C. R. Acad. Sci. Ser. A296 881 (1983).

7.   D. Ben-Avraham and S. Havlin, J. Phys. A15, L691 (1982).

8.   J. Klafter, A. Blumen and G. Zumofen, J. Stat. Phys. 36, 561 (1984).

9.   P. Argyrakis and R. Kopelman, Phys. Rev. B29, 511 (1984).

10. D.C. Ahlgren, Ph.D. Thesis, University of Michigan (1979).

11. J.S. Newhouse, Ph.D. Thesis, University of Michigan (1985).

12. R. Kopelman, P.W. Klymko, J.S. Newhouse and L.W. Anacker, Phys. Rev. B29, 2164 (1984).

13. L.W. Anacker and R. Kopelman, J. Chem. Phys. 81, 6402 (1984).

14. R. Kopelman, E.M. Monberg, F.W. Ochs and P.N. Prasad, J. Chem. Phys. 62, 292 (1975).

15. R. Kopelman, E.M. Monberg, F.W. Ochs and P.N. Prasad, Phys. Rev. Lett. 34, 1506 (1975).

16. R. Kopelman, Topics in Applied Physics 15, 247 (1976).

17. J. Hoshen and R. Kopelman, J. Chem. Phys. 65, 2817 (1976).

18.   P. Argyrakis, Ph.D. Thesis, The University of Michigan (1978).

19.   P.C. de Gennes, La Recherache, 7, 919 (1976).

20.   R.B. Pike and H.E. Stanley, J. Phys. A14, L169 (1981).

21.   R. Kopelman and P. Argyrakis, J. Chem. Phys. 72, 3053 (1980).

22.   P.W. Klymko and R. Kopelman, J. Lumin. 24/25, 457 (1981).

23.   P.W. Klymko and R. Kopelman, J. Phys. Chem. 86, 3686 (1982).

24.   E.W. Montroll and H. Scher, J. Stat. Phys. 9, 101 (1973).

25.   H. Scher and M. Lax, Phys. Rev. B7, 449 (1973).

26.   H. Scher and E.W. Montroll, Phys. Rev. B12, 2455 (1975).

27.   M.F. Shlesinger, J. Stat. Phys. 10, 421 (1974).

28.   M.F. Shlesinger, J. Stat. Phys. 36, 639 (1984).

29.   L.W. Anacker, R. Kopelman and J.S. Newhouse, J. Stat. Phys. 36, 591 (1984).

30.   L.W. Anacker and R. Kopelman, Phys. Rev. B (in press).

31.   L.W. Anacker and R. Kopelman, this volume.

32.   J. Prasad, S.J. Parus and R. Kopelman, this volume.

33.   E.I. Newhouse and R. Kopelman, this volume.

34.   L.A. Harmon and R. Kopelman, this volume.

35.   L.W. Anacker, P.W. Klymko and R. Kopelman, J. Lumin. 31/32, 648 (1984).

36.   L.A. Harmon, Ph.D. Thesis, The University of Michigan (1985).

37.   U. Even, K. Rademann, J. Jortner, N. Manor and R. Reisfeld, Phys. Rev. Lett. 42, 2164 (1984).

38.   P.W. Klymko, Ph.D. Thesis, The University of Michigan (1984).

39.   A.H. Francis and R. Kopelman, Topics in Applied Physics 49, 241 (1981).

40.   D. Stauffer, Introduction to Percolation Theory, Taylor and Francis, London (1985).

41.   J.S. Newhouse, J. Hoshen and R. Kopelman (unpublished).

FRACTAL-LIKE EXCITON KINETICS IN POROUS GLASSES, MEMBRANES AND POWDERS

J. Prasad, S. Parus and R. Kopelman

Department of Chemistry
The University of Michigan
Ann Arbor, Michigan 48109

ABSTRACT

We have measured the exciton (triplet) recombination (fusion, anni-hilation) characteristics of naphthalene-doped microporous materials. This technique yields the dynamic (spectral, fracton) dimension of the embedded naphthalene structure or the effective random walk dimension of the porous network. Temperature studies separate the energetic and geometric features of the pore space. The geometric dynamic (spectral) dimensions are mostly between 1 and 2, i.e. fractal-like, and are consistent with previous results on the static (fractal or Euclidean) dimensions of the porous vycor glass samples.

INTRODUCTION

Porous glasses, rocks and related materials have commanded much interest lately and the fractal dimension of the pores has been of much discussion.[1-4] Anomalous diffusion and the associated fracton or spec-tral dimension[5-7] have become a new tool in the characterization of fractal properties.[8-11] However, while the diffusion experiments require microscopic measurements, often below the optical diffraction limit,[8] the anomalous reaction kinetics, which are a direct consequence of anoma-lous diffusion,[12-18] can be studied via macroscopic measurements. This has been demonstrated by simulations[11-16,19] as well as by exciton anni-hilation studies on isotopic alloy crystals[11,17,18] (where the exciton transport is confined to the percolation clusters of the lower excitation energy component). In these isotopic alloys an energetic restriction leads to clear-cut fractal spaces (incipient infinite percolation clusters) with a measured fracton (spectral) dimension,[21,22] d_s, of about 4/3, while the steady-state energy transport experiments[23,24] give the frac-tal dimension, d_f (within three significant figures).

In this work we demonstrate that a fractal-like triplet exciton fusion kinetics does occur in several classes of disordered media, of both intrinsic and practical interest: porous glasses, porous membranes, and filter papers. In all these matrices the exciton conductor is naphthalene (which is embedded into the porous matrices). The naphthalene domains appear to have a fractal geometry. The question of interest is whether such geometry is due only to trivial considerations (the glass wall) or whether there are also more subtle effects (e.g., naphthalene domain-boundaries of fractal shape). Our experiments do unscramble simple geometric effects from energetic effects.

SAMPLES AND THEIR PREPARATION

The sample preparation of the porous polymeric membranes and filter papers involved soaking in solutions of naphthalene (zone refined, in spectroscopic grade hexane). The porous vycor (7930 Corning glass) samples were prepared by naphthalene sublimation in vacuum. The original vycor glass was treated with acids, water and alcohol. The samples were excited at low temperatures by a 1600 W xenon arc lamp through a monochromator set at 310 nm. Spectra were collected via a 1 meter JY double monochromator (with EMI 9816 QB PMT and PAR 1109 photon counter). The delayed fluorescence decays were obtained by shuttering the excitation beam and using various filters with an EMI 978IR phototube and a PAR 4202 signal averager.

EXCITON KINETICS

The principle behind the approach taken here is the relation of the instantaneous exciton annihilation probability to the site visitation efficiency. The instantaneous exciton annihilation is $d\rho/dt$ where $\rho(t)$ is the exciton density. The fraction of exciton population annihilated (in time dt) is $\rho^{-1}d\rho/dt$. This fraction is linear in density ρ (binary collision theory argument). Hence the normalized instantaneous exciton annihilation probability k is given by:

$$k = \rho^{-2}d\rho/dt \ . \tag{1}$$

Classically k is time independent ("rate constant" in chemical language). It has been shown that k is proportional to the visitation efficiency ε of a random walker[12,13,17](the time derivative of the mean number of <u>distinct</u> sites visted):

$$k \sim \varepsilon \sim t^{-h} \qquad\qquad 0 \leq h \leq 1 \tag{2}$$

where h = 0 for 3-dimensional Euclidean spaces and lattices (classical result) but for $d_s < 2$, e.g. fractal domains,[17]

$$h = 1 - d_s/2 \ . \qquad\qquad h > 0 \tag{3}$$

The delayed fluorescence (F) feeds directly on the triplet exciton fusion: $F(t) \sim d\rho/dt$. The phosphorescence (P) is due to the natural decay of the triplet exciton population: $P(t) \sim \rho(t)$. From Eqs. 1, 2 we get:

$$F/P^2 \sim t^{-h} \ . \qquad\qquad 0 \leq h \leq 1 \tag{4}$$

We note that under special circumstances, where one of the two fusing excitons is trapped ("heterofusion"), and is present in large abundance (relative to the freely moving excitons), one gets,[17] instead of the binary reaction (1) ("homofusion"), a pseudo-unary reaction:

$$k = \rho^{-1}d\rho/dt \tag{5}$$

and thus

$$F/P \sim t^{-h} \ . \tag{6}$$

We note that for both (4) and (6) one has at early times (P ~ ρ ≈ const):

$$F \sim t^{-h} \, . \qquad\qquad t \to 0 \qquad\qquad (7)$$

SPECTRAL STUDIES

The fluorescence and phosphorescence spectra show broadened bands, physical trap emissions, excimer emission and radical (1-hydronaphthyl) emission. All these are characteristic features of disordered aromatic solids. Spectra of naphthalene in vycor depend on temperature. In the membrane and filter paper samples the temperature does not affect the emission spectra, thus making these samples easier to study. A weak phosphorescence intensity is common to all samples. This may be, in part, because of the triplet-triplet annihilation (fusion). See Figs. 1, 2.

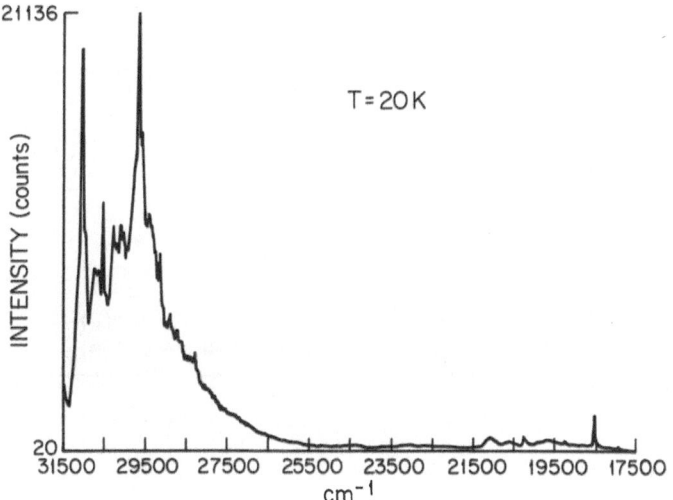

Fig. 1. Emission spectra of naphthalene embedded in nylon membrane.

RESULTS

Our experimental results are shown in Figs. 3 and 4. The heterogeneity exponent h is extracted from the slopes of ℓn k vs ℓn t. The values so obtained for h are listed in Table 1, along with values for the

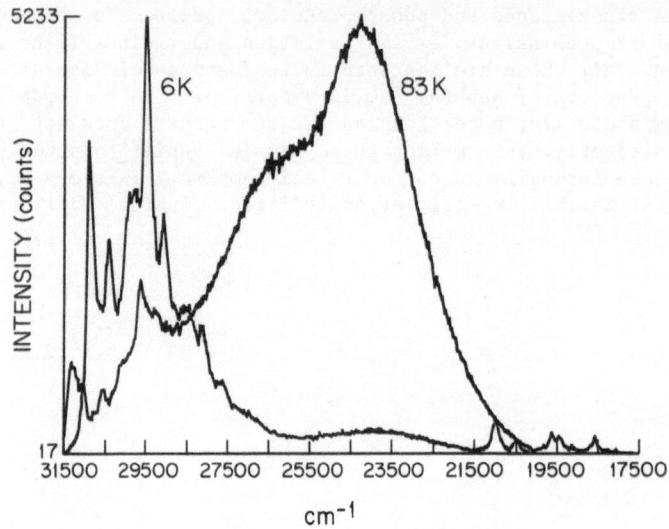

Fig. 2. Emission spectra of naphthalene embedded in vycor at 6K and 83K.

Table 1

Sample	Pore Size(μ)	T (K)	Exponent(h) Homofusion	Spectral dim.($d_s^!$)
Acetate (GA8) Membrane[†]	0.2	4	0.16	1.7
		80	0	3
Acetate (GA1) Membrane[†]	0.5	4	0.47	1.1
Acetate (GA3) Membrane[†]	1.2	4	0.44	1.1
Nylon (B214) Membrane[†]	0.2	4	0.21	1.6
Glass Filter Paper	0.6	6	0.42	1.2
		80	0.20	1.6
Cellulose Filter Paper	0.6	6	0.33	1.3
		80	0.06*	1.9 – 3
Vycor Porous Glass	0.003	6	0.44	1.1
		58	0.5	1

* Heterofusion (a homofusion model gives a negative value)
† Gelman Sciences (courtesy of A. Korin and M. Kraus)

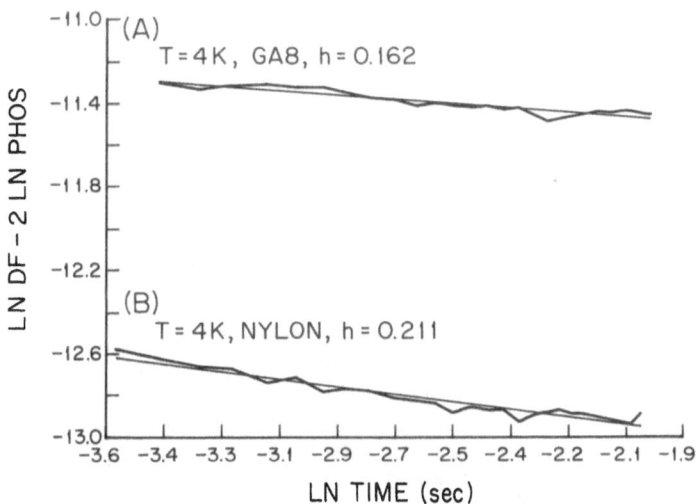

Fig. 3. Annihilation rate coeff., k = (DF)/(Phos) vs. t on a ln-ln plot for polymeric membranes. H is the heterogeneity exponent h.

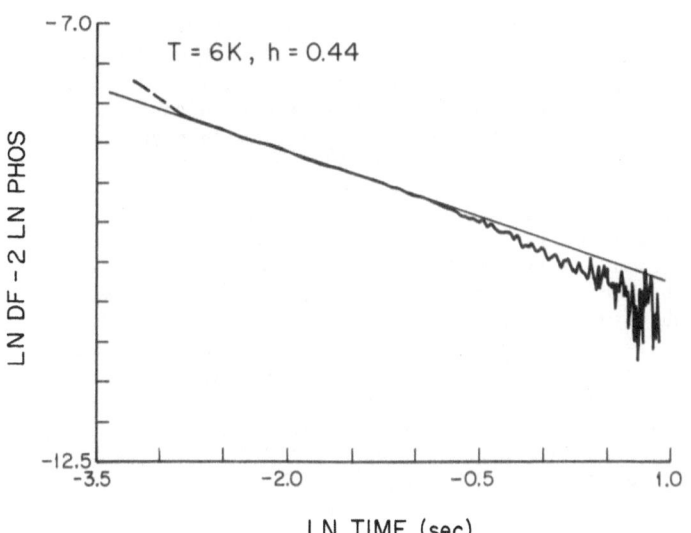

Fig. 4. Same type of plot as in Fig. 3 for vycor porous glass.

effective spectral dimension $d_s^!$ of the fractal-like medium calculated from the relation $d_s^! = 2(1 - h)$. We note that for all samples at low temperatures $h > 0$. This implies that all samples exhibit a fractal-like behavior at these low temperatures. We also note that for the vycor sample $d_s^! \approx 1.1$. This value does not change much with increasing temperature. On the other hand, we observe (Table 1) some drastic temperature effects on the membranes and filter paper values; notably the cellulose filter paper approaches a classical behavior at $T \approx 80K$, i.e., $h \to 0$.

DISCUSSION

We have shown experimentally, using vapor-deposited naphthalene,[9,25] as well as by simulations,[26] that random energetic disorder can also result in fractal-like kinetics (and thus in an effective h and d_s). Moreover, geometric and energetic effects do superimpose[27] in a way that is analogous to the "subordination principle" of Klafter, Blumen and Zumofen.[2] In the latter, the spectral dimension d_s is effectively reduced by a constant factor accounting for a "continuous time random walk" (CTRW) with an anomalous hopping time distribution (infinite second moment). We note that this CTRW model has been shown to be equivalent to models of energetic disorder.[2] We can thus separate the roles of geometric and energetic disorder via temperature studies. In the limit where the thermal energy is large compared to the inhomogeneous line-broadening (which we observe spectroscopically), the effects of the energetic disorder are minor and the geometric effect dominates. Table 1 shows that both in glass (synthetic) and cellulose (natural) <u>filter papers</u> the triplet excitons of the naphthalene embedded in the pores move anomalously (at low temperatures) due to energetic disorder. However, it appears that while the glass filter paper has a pore network with a real fractal-like structure (geometric spectral dimension of 1.6) the natural filter paper has $h \approx 0$, which is consistent with $d_s \geq 2$, i.e. a non-fractal (Euclidean) pore network (which would be more efficient for biological diffusion processes).

The low spectral dimension value ($d_s = 1.1 \pm 0.2$) for our sample of porous vycor (which supposedly is the same as that used by Dozier et al.[8]) is consistent with a number of literature claims. Even et al.[1] argued that $d_f = 1.7$ for their porous vycor sample (not necessarily the same as ours) and interpreted it in terms of a percolation backbone. This would result in $1 < d_s \leq 4/3$. Dozier et al.[8] have chosen to model their porous glass sample with a 3-dimensional percolation cluster ($d_f \approx 2.5$). This gives $d_s \approx 1.3$. Yang et al. quote a totally different model (for an unspecified sample) consisting of a simple cubic lattice made of interconnecting spheres and cylinders. If the cylinders are narrow enough, they (literally) become the "bottle-neck" for the exciton recombination reaction and thus result in a one-dimensional effective topology ($d_s = 1$). [We note that we have also modelled[28] a similar quasi-one-dimensional network (parallel connected lines) and got $d_s \approx 1$ over the appropriate time scale]. Our experimental values are thus quite believable but not yet very discriminating (a 10% error in h translates into a 20% error in d_s). We note that for any connected fractal structure, $1 \leq d_s \leq 2$. [A fractal "dust" will result in $d_s < 1$ ($h > 0.5$); however, the annihilation efficiency on a "dust" structure is quite low, resulting in extremely low delayed fluorescence signals.]

The dynamic range of our present measurements are limited by a 5 ms shutter resolution (which can be improved) and the 2-3 second triplet lifetime (limiting signal/noise). We estimate that the corresponding length-scale for exciton migration is on the order of 1000 Å (0.1 μ). We note that the nominal pore sizes (Table 1) give upper limits, rather than mean values, for the membranes and filter papers. Due to the much faster exciton migration in Euclidean (ordered) spaces (the bulk migration is at least two orders of magnitude more efficient than migration in fractal-like domains) we believe that our data on these samples characterize the more numerous but narrower microporous networks, rather than the nominal pore sizes. The only exception is the porous vycor glass sample whose pore size is a mean value based on molecular adsorption measurements.[8]

POWDERED NAPHTHALENE SAMPLES

Insight into effects of the embedding medium may be obtained by comparison to unsupported powdered naphthalene. At 27K, lifetimes of delayed fluorescence and delayed excimer emission are comparable. At 6K, the delayed fluorescence lifetime is much shorter than the delayed excimer lifetime. This contrasts with vycor and membrane samples where the rates of the two delayed emissions are similar at both temperatures. In the latter types of environments, the ordered and disordered regions may be intermixed. The excitons in the ordered regions could transport to the disordered regions where the excimers exist. Thus the normally slow transport in the disordered regions would be enhanced by the rapid transport in the ordered regions. In a powder, disordered regions are expected only on the outer surface of a crystalline particle. Transport is rapid in the inner portion where delayed emission is consequently observed as fluorescence. The slower transport occurs on the outside where the disorder results in the emission being from the excimers. At 27K, the rate of transport in the disordered region is thermally assisted and thus accelerated. The pore size studies might be helpful in characterizing the crystalline domain sizes (providing upper limits for the latter).

CONCLUSIONS

In conclusion, the studies of naphthalene embedded in polymeric membranes, filter papers and vycor glass samples have demonstrated a fractal-like exciton annihilation kinetics. Temperature studies establish both the subordination of geometric and energetic disorder and a technique for assessing the relative roles of geometric and energetic constrains. The network spectral dimension is usually in the range $1 < d_s < 2$, but some samples are consistent with $d_s = 1$ and others with $d_s = 3$. Pore size studies of polymeric membranes might characterize the crystalline domain sizes in deposited films.

ACKNOWLEDGEMENTS

This work was supported by NSF Grant No. DMR 8303919.

REFERENCES

1. U. Even, K. Rademann, J. Jortner, N. Manor and R. Reisfeld, Phys. Rev. Lett. 42:2164 (1984).
2. J. Klafter, A. Blumen and G. Zumofen, J. Stat. Phys. 36:561 (1984).
3. C. L. Yang, P. Evesque and M. A. El-Sayed, J. Phys. Chem. 89:3442 (1985).
4. A. J. Katz and A. H. Thompson, Phys. Rev. Lett. 54:1325 (1985).
5. Y. Gefen, A. Aharony and S. Alexander, Phys. Rev. Lett. 50:77 (1983).
6. S. Alexander and R. Orbach, J. Phys. (Paris) Lett. 54:L625 (1982).
7. R. Rammal and G. Toulouse, J. Phys. (Paris) Lett. L13 (1983).
8. W. D. Dozier, J. M. Drake and J. Klafter, Phys. Rev. Lett. (Jan. 86).

9. L. A. Harmon and R. Kopelman, J. Lumin. 31/32:660 (1984).

10. E. I. Newhouse and R. Kopelman, J. Lumin. 31/32:651 (1984).

11. R. Kopelman, J. Stat. Phys. 42:185 (1986).

12. R. Kopelman and P. Argyrakis, J. Chem. Phys. 72:3053 (1980).

13. P. G. de Gennes, C. R. Acad. Sci. Ser. A296:881 (1983).

14. I. Webman, Phys. Rev. Lett. 52:220 (1984).

15. K. Kang and S. Redner, Phys. Rev. Lett. 52:955 (1984).

16. P. Meakin and H. E. Stanley, J. Phys. A 17:L173 (1984).

17. P. W. Klymko and R. Kopelman, J. Phys. Chem. 87:4565 (1984).

18. P. Evesque and J. Duran, J. Chem. Phys. 80:3016 (1984).

19. J. S. Newhouse, P. Argyrakis and R. Kopelman, Chem. Phys. Lett. 107:48 (1984).

20. A Aharony and D. Stauffer, Phys. Rev. Lett. 52:2368 (1984).

21. R. Kopelman, P. W. Klymko, J. S. Newhouse and L. W. Anacker, Phys. Rev. B 29:3747 (1984).

22. L. W. Anacker and R. Kopelman, J. Chem. Phys. 81:6402 (1984).

23. D. C. Ahlgren, Ph.D. Thesis, University of Michigan (1979).

24. J. S. Newhouse, Ph.D. Thesis, University of Michigan (1985).

25. L. A. Harmon, Ph.D. Thesis, University of Michigan (1985).

26. L. W. Anacker, R. Kopelman and J. S. Newhouse, J. Stat. Phys. 36:591 (1984).

27. L. W. Anacker and R. Kopelman, Phys. Rev. B (in press).

28. L. W. Anacker, L. Lee and R. Kopelman (unpublished).

REACTION KINETICS ON FRACTAL AND EUCLIDEAN STRUCTURES WITH ENERGETIC

DISORDER: RANDOM WALK SIMULATIONS

L. W. Anacker and R. Kopelman

Department of Chemistry
The University of Michigan
Ann Arbor, Michigan

ABSTRACT

A Monte Carlo approach is used to treat diffusion-limited reaction
kinetics in microscopically heterogeneous media with energetic disorder.
Simulations performed on the Cyber 205 supercomputer for the elementary
reaction A + A → A show that the rate law is well described in terms of
the microscopic exploration space of a single random walker, that is, by
the number of distinct sites visited, S. This provides a scaling approach
for moving from single walker simulations to reacting random walker simulations over a broad range of times and reduced temperatures. In the
asymptotic limit of long times, single walker simulations for exponential,
Gaussian and uniform distributions of energetic disorder on the Sierpinski
gasket appear to follow a simple power law, $S(t) \propto t^f$. Simulations of reacting random walkers show that the density, ρ, is fairly well described
by the relation: $\rho^{-1} \propto S(t)$ as $t \to \infty$.

INTRODUCTION

Diffusion-limited reaction kinetics in disordered and low-dimensional
media have non-classical rate laws with strongly time-dependent rate coefficients [1-16]. Transport as well as diffusion-limited chemical reactions exhibit anomalous time-dependent behavior in the presence of fractal
geometries and/or spatial [1-12,16-22], temporal [23-29], or energetic
[7,25,30-35] disorder. As a first order approximation, a Smoluchowski-
type stochastic approach [12,36-38] for the A + A → A reaction is adopted
which relates the microscopic transport rate (measured as the number of
distinct sites visited in time t by a single walker, S(t)) to the macroscopic rate of decay. The diffusion-controlled kinetics for elementary,
binary reactions A + A → A are well described [7-9,12] by:

$$- \frac{d\rho}{dt} = k_0 \left(\frac{dS}{dt}\right) \rho^2 .$$

(1)

Here we focus on the range of validity of our stochastic approach using
large-scale Monte Carlo simulations on fractal Sierpinski gaskets and
Euclidean lattices; both systems include the superposition of energetic
disorder. This Smoluchowski-type approach does quite well [12] in the
absence of disorder for the Sierpinski gaskets in dimensions two through
five and for Euclidean lattices in one, two and three dimensions.

Results reported here present a more stringent test of the Smoluchowski-type approach arising from the additional constraint of energetic disorder.

Historically, the Smoluchowski approach involved the diffusion constant rather than the number of distinct sites visited. The Rosenstock [39] approximation first linked the number of distinct sites visited, $S(t)$, to the probability of trapping for reactions $A + B \to B$ where A are mobile and B are stationary traps. This was generalized [38] to include disordered media and resulted in the time-dependent rate coefficient, $k(t) \propto dS/dt$. The rate coefficient for the reaction $A + A \to$ Products was conjectured [4c] to have the same functional form when some A are stationary and others are mobile. It was further implied [1,4c,40] that the same conjecture might hold for the case where all A are mobile even though this is a more subtle case; this has been proven for the one-dimensional case [14-16]. Extensive simulations [6-13] and experiments [2-7] for the reaction $A + A \to A$ substantiate the validity of this conjecture: $k(t) \propto dS/dt$.

MODEL

Systems with fractal geometries are well characterized by their Euclidean dimension, d, fractal dimension [41], d_f, and spectral dimension [17,18], d_s. Systems with spatial and energetic disorder are related to those with temporal disorder [23-25]: "All the randomness of space is incorporated into choosing the appropriate waiting-time distribution " [24], $\psi(t)$. This type of temporal disorder is often referred to as continuous time random walk [23,26-29]. A fractal distribution of event times [24] is generated when the mean time between events is infinite while the median time is finite; this occurs [23-25] for any distribution $\psi(t)$ with a long-time tail of the form $t^{-1-\lambda}$ where $0<\lambda<1$. The superposition of temporal disorder on a system with spatial disorder or on fractal structures [23] requires d, d_f, d_s, and λ to characterize the dynamics. Energetic disorder, as considered here, introduces a temperature-dependent fractal-like distribution of waiting times with a Boltzmann-weighted probability for moves to sites of higher energy [9,22]. A walker moves from site i to one of its z nearest-neighbor sites with probability:

$$
P_{ij} = \begin{cases} z^{-1}\exp(-(\epsilon_j - \epsilon_i)/kT) & \epsilon_j > \epsilon_i \\ z^{-1} & \epsilon_j \leq \epsilon_i \end{cases} \tag{2a}
$$

or remains on site i with probability:

$$
P_{ii} = 1 - \sum_{j=1}^{z} P_{ij}. \tag{2b}
$$

Note the asymmetries: $P_{ij} \neq P_{ji}$ and therefore, unlike the continuous time random walk [23,26-29], $<t_{ij}> \neq <t_{ji}>$.

Transport is measured by the mean number of distinct sites visited, $<S(t)>$, at time t. For several types of disorder $<S(t)>$ is surprisingly well described by a simple power law [18-24]:

$$
<S(t)> \propto t^f \qquad t \to \infty \tag{3}
$$

in which:

30

$$f = \begin{cases} 1 & d_S > 2 & \text{(spatial)} \\ d_S/2 & d_S < 2 & \text{(spatial)} \\ \lambda & 0 < \lambda < 1 & \text{(temporal)} \\ \lambda d_S/2 & d_S < 2 & \text{(spatial + temporal)} \\ f'(T) & & \text{(energetic)} \end{cases}$$

Higher order terms exist, but for the most part do not contribute significantly. Figs. (1a), (1b) and (1c) illustrate the effect of temperature on $S(t)$ for the square lattice with a uniform pseudo-random distribution

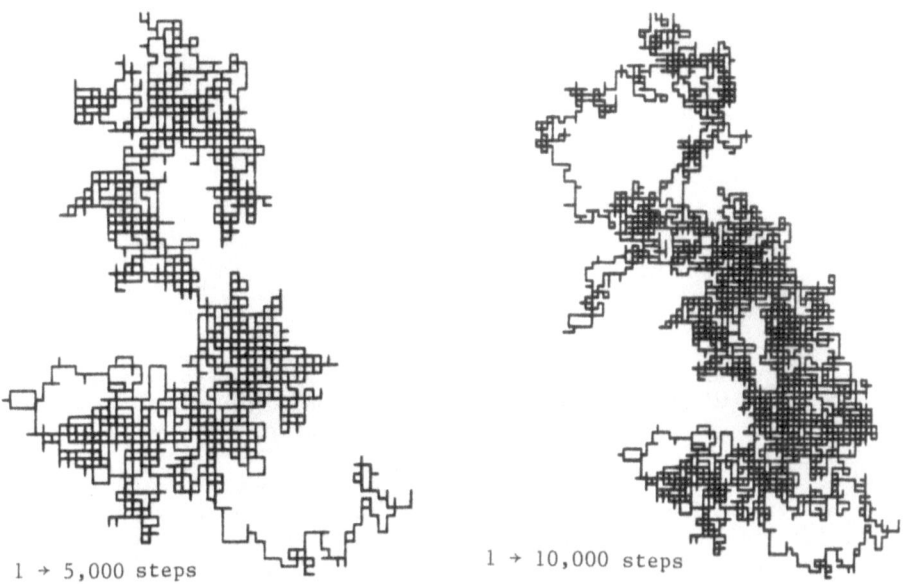

1 → 5,000 steps   1 → 10,000 steps

Fig. 1a: The trail mapped out after 5,000 and 10,000 steps by a single random walker on a square lattice with uniformly distributed random site energies at a reduced temperature of T" = 1.0.

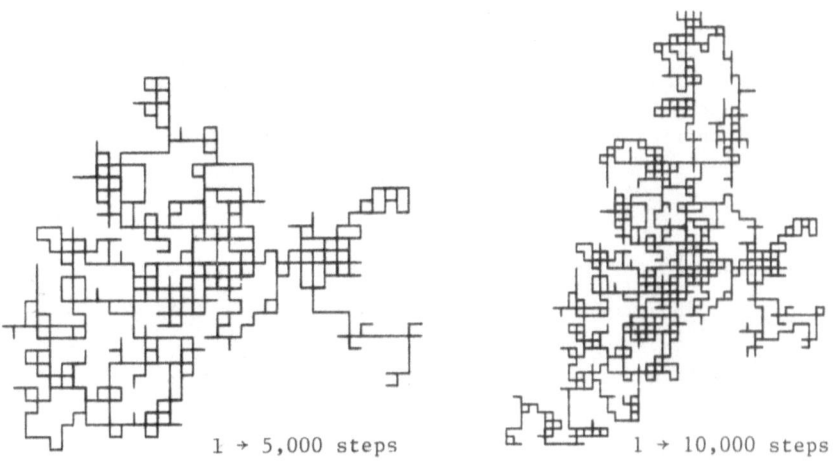

1 → 5,000 steps   1 → 10,000 steps

Fig. 1b: The trail mapped out after 5,000 and 10,000 steps by a single random walker on a square lattice with uniformly distributed random site energies at a reduced temperature of T" = 0.2.

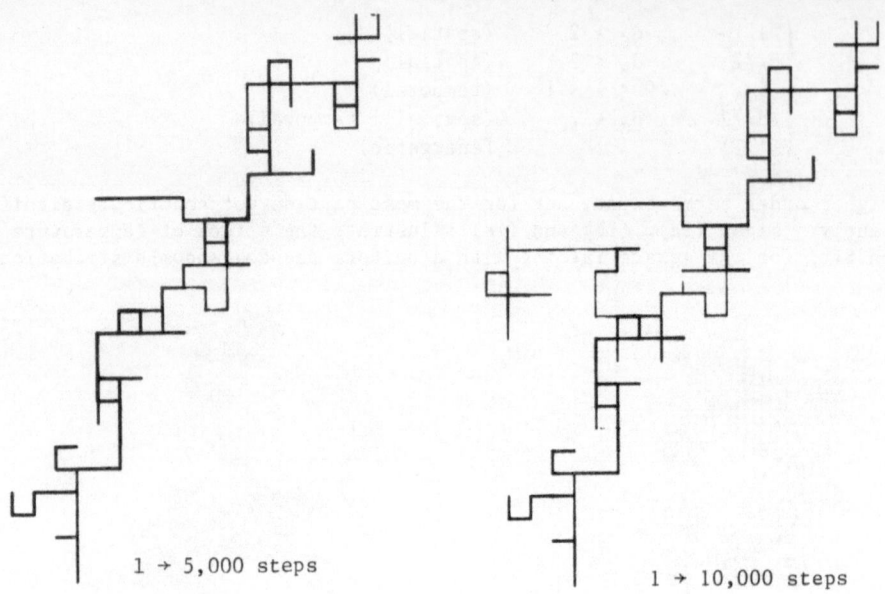

1 → 5,000 steps   1 → 10,000 steps

Fig. 1c: The trail mapped out after 5,000 and 10,000 steps by a single
          random walker on a square lattice with uniformly distributed
          random site energies at a reduced temperature of $T'' = 0.1$.

of site energies. The highest temperature, Fig. (1a), has very good transport and $S(t)$ is almost linear with time. This is contrasted with the
lowest temperature, Fig. (1c), at which only ≈ 20 new sites are visted in
the last 5,000 steps. In the section on Transport on the Energetically
Disordered Gasket, the effects of various distributions of site energies
(exponential, Gaussian and uniform) are presented for single walker simulations on the planar Sierpinski gasket. For each distribution, Eq. (3)
is a fairly good approximation over a broad range of reduced temperatures.

Consider now diffusion-limited chemical reactions with a bimolecular
decay mechanism for the A + A reaction. When two particles meet they
react instantaneously, and the rate of reaction is governed strictly by
the rate of transport in the system considered. Experiments [2-6,30,33]
and extensive random walk simulations [4-13] of systems with spatial,
temporal, or energetic disorder exhibit anomalous time-dependent behavior
for the rate coefficient of the second-order rate law describing the
A + A reaction:

$$- \frac{d\rho}{dt} = k(t)\rho^2. \tag{4}$$

This corresponds to the Smoluchowski-type equation:

$$- \frac{d\rho}{dt} = k_0 \left(\frac{dS}{dt}\right) \rho^2 \tag{5}$$

which can be written as:

$$\frac{d\rho^{-1}}{dt} = - \rho^{-2} \frac{d\rho}{dt} = k_0 \left(\frac{dS}{dt}\right) \tag{6}$$

or integrated to obtain:

$$\rho^{-1} - \rho_0^{-1} = k_0(S(t) - S_0) \qquad (7)$$

where $\rho_0 \equiv \rho(t=0)$ and $S_0 \equiv S(t=0)$. For times long enough so that the initial conditions can be ignored, Eqs. (3) and (7) can be combined for the A + A reaction to give:

$$<\rho^{-1}> \propto <S(t)> \propto t^f \qquad (8)$$

which is consistent with earlier results [4,9-13] for this reaction. In the section on the Smoluchowski-type Approach, we test the range of validity of Eq. (7) using large-scale Monte Carlo simulations of single walkers and reacting random walkers on the energetically disordered planar Sierpinski gasket, square lattice, and simple cubic lattice.

SIMULATION METHODOLOGY

Random walk simulations were used to study transport and diffusion-limited chemical reactions on 8th order planar Sierpinski gaskets (9843 sites), and on square (512^2 sites) and simple cubic (64^3 sites) lattices. All simulations on Euclidean lattices imposed periodic boundary conditions while the three vertices of the largest triangle on the Sierpinski gasket formed reflective boundaries. Lattice site energies for each run were assigned when a walker tried to visit a site with a previously unassigned site energy; once a site energy was assigned it retained that fixed value. The direction of motion for every walker at each time increment was decided by a uniform pseudo-random number generator; time was incremented after all walkers attempted to move. A new lattice was generated for each run.

Initial conditions for single walker simulations set $S_0 \equiv S(N=0) = 1$. Single walker simulations on the Sierpinski gasket started each random walker at a random initial site; on Euclidean lattices, the walker started at the center of the lattice. Reacting random walker simulations set $\rho_0 \equiv \rho(N=0) = 0.1$ with an initial spatial distribution of walkers assigned using a uniform pseudo-random number generator.

Unitless temperature parameters were used for each energy distribution: $T'' = kT/W$ with a maximum allowed energy difference of W. Exponential site energies were assigned from a normalized distribution with mean 1; discrete energies were used in such a way that energies below 0.001 were not distinguished from energies of 0.001 and energies above 10 were assigned an energy of 10, i.e. $W = 10$. A normalized Gaussian distribution was used with mean 0, standard deviation $\sigma = 1$, and an energy cut-off of 3σ, i.e. $W = 6\sigma$; energies outside the range $(-3\sigma, 3\sigma)$ were included but they were not distinguished from energies of $\pm 3\sigma$. The uniform energetic distribution, $T'' = kT/W$, referred to here corresponds to the random site energies being uniformly distributed from 0 to 1: $W = 1$. Pseudo-random number generators used were: RANF on the CYBER 205 at Colorado State University for uniform distributions; and FUNIF, FNORM and FEXP on the Amdahl 5860 at The University of Michigan for the uniform, Gaussian and exponential distributions, respectively.

Results reported in the next section strictly concern single walker transport on the planar Sierpinski gasket. Three types of pseudo-random energetic distributions were considered: exponential, Gaussian and uniform. These simulations consisted of 1,000 realizations for each reduced temperature shown in Table 1.

Table 1. Unitless Temperature (T") Values for Simulations of Single
Walker on a Planar Sierpinski Gasket*

Uniform:	T" = .10,	.15,	.25,	.50,	1.					
Gaussian:	T" = .03,	.05,	.06,	.07,	.08,	.09,	.1,	.15,	.2,	1.
Exponential:	T" = .01,	.03,	.05,	.10,	.20,	1.				

*1000 realizations at each T"; 4000 time increments.

TRANSPORT ON THE ENERGETICALLY DISORDERED GASKET

Log-log plots of $<S(N)>$ versus N for exponential, Gaussian and uniform
distributions of energetic disorder are shown in Fig. (2). In each case,
the curves become fairly linear for large N in good agreement with Eq. (3).
At high temperatures, $f(T") = d_s/2$ as expected. Lower temperatures illustrate effects of energetic disorder on a fractal; this is comparable to the

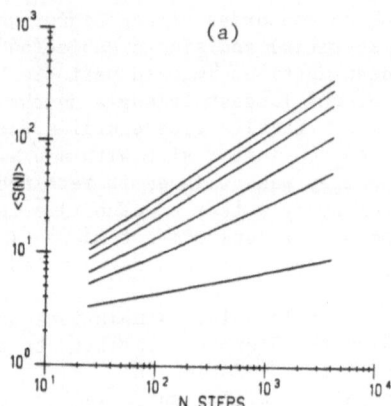

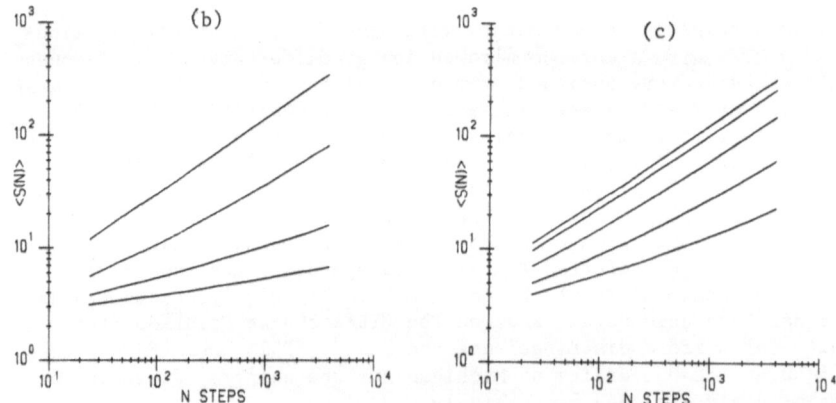

Fig. 2. Mean number of distinct sites visited, $<S(N)>$, by a single random walker on an 8th order planar Sierpinski gasket with a) exponential [T" = .01, .03, .05, .1, .2, 1.0], b) Gaussian [T" =
.03, .05, .1, 1.0] and c) uniform [T" = .1, .15, .25, .5, 1.0]
distributions of random site energies. Bottom to top curves
correspond to increasing T".

superposition of temporal disorder on a fractal ("subordination") [23,24].
The effect of subordination is most clearly seen in Fig. (2a), the exponentially distributed site energies, where each curve becomes linear in the
limit of large N. Similar behavior is seen qualitatively for the Gaussian
and uniform distributions, Figs. (2b) and (2c), respectively; however, these
distributions may be more sensitive to the asymmetric hopping probabilities.

Figs. (2b) and (2c) clearly demonstrate nonlinear behavior at lower
temperatures. At early times the curves have slopes which correspond to
$f(T'') < d_s/2$, but for larger N the beginnings of what may be a crossover
to fractal behavior $f(T'') \rightarrow d_s/2$ are seen. Note that a self-similar fractal distribution of waiting times is not expected to exhibit crossover behavior. Whether the powers $f(T'')$, from the exponential distribution of site
energies, are still approaching $d_s/2$ at lower temperatures is not clear.
[Future research into "self-affine fractals" [41] may help explain this
crossover regime.] For all three distributions of site energies, the functional dependence of $f(T'')$ on the reduced temperature is nonlinear as illustrated in Fig. (3); this will be discussed in greater detail in a separate publication.

The next section combines results from single random walker simulations and reacting walker simulations. Reacting walker simulations allowed
only one walker to occupy each site at any given instant. Walkers were
so placed on the lattice that each walker had an equal probability of
landing on any unoccupied site. When a walker tried to move to a site
already occupied, it was removed without further perturbing the system,
i.e., $A + A \rightarrow A$. For the isoenergetic perfect ($P_{ij} = P_{ji}$) Sierpinski Gasket and for each T'' shown in Table 2, simulations consisted of 100 to 1000
realizations for single random walkers and 10 to 20 runs for reacting random walkers.

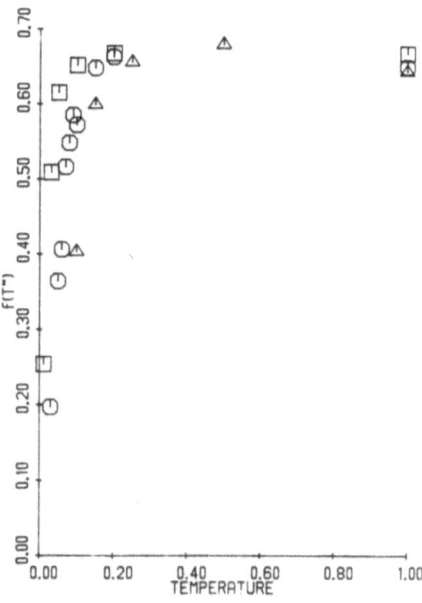

Fig. 3. $f(T'')$, Eq. (3), as a function of the temperature, T'', for the exponential (squares), Gaussian (circles), and uniform (triangles)
distributions of random site energies on a planar Sierpinski
gasket. Nonlinear regressions in the long-time limit were used
to obtain these values of $f(T'')$.

Table 2. Unitless Temperature (T") Values for Systems with Both
          Single Walker and Reacting Walker Simulations

Square Lattice*
   Uniform: T" = 0.05, 0.06, 0.07, 0.08, 0.09, 0.1

Planar Sierpinski Gasket**
   Uniform: T" = 0.10, 0.15, 0.20

Simple Cubic Lattice*
   Uniform: T" = 0.05, 0.055, 0.06, 0.065, 0.07, 0.1
   Gaussian: T" = 0.05, 0.066, 0.083, 0.166

---

* 10,000 time steps
** 4,000 time steps

For the simple cubic lattice, reacting walker simulations on the perfect lattice included 120 realizations. Reacting walker simulations with
a Gaussian distribution of site energies were performed on a smaller scale;
10 runs were performed for each T", and these simulations initially placed
1563 walkers on a 25^3 site lattice.

SMOLUCHOWSKI-TYPE APPROACH

In Fig. (4), $\langle \rho^{-1} - \rho_0^{-1} \rangle$ is plotted versus $\langle S - S_0 \rangle$ for the simulations on the simple cubic lattice with uniform site energies. The curves
are found to be roughly linear as expected from Eq. (7). To better

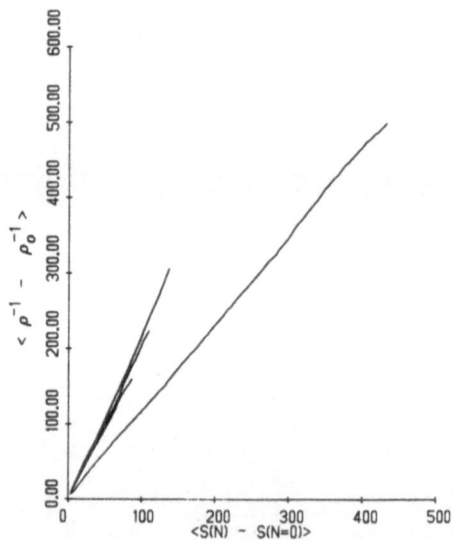

Fig. 4. $\langle \rho^{-1} - \rho_0^{-1} \rangle$ versus $\langle S(N) - S(N=0) \rangle$ for the simple cubic lattice
          with a uniform distribution of random site energies. Shortest
          to longest curves correspond to increasing temperatures: T" =
          0.06, 0.065, 0.07, and 0.1.

illustrate the simulation results, Eq. (7) is written as:

$$\frac{\langle \rho^{-1} - \rho_0^{-1} \rangle}{\langle S(t) - S_0 \rangle} = k_0 \qquad (9)$$

A plot of Eq. (9) as a function of time should generate a horizontal curve, slope = 0, if Eqs. (5) through (7) are exact. The validity of

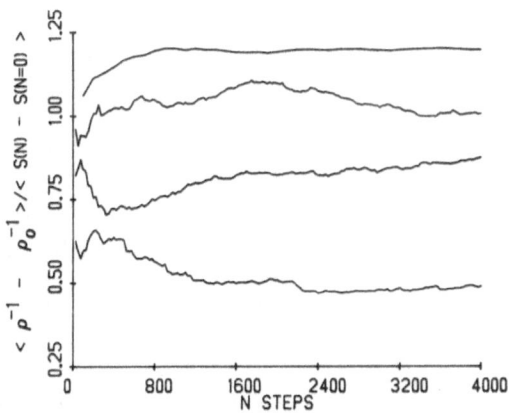

Fig. 5. Left hand side of Eq. (9) versus time for the planar Sierpinski Gasket with uniformly distributed site energies for (bottom to top): $T'' = 0.1$, 0.15, and 0.2, and the isoenergetic Sierpinski Gasket. Curves shifted along y-axis, for illustration.

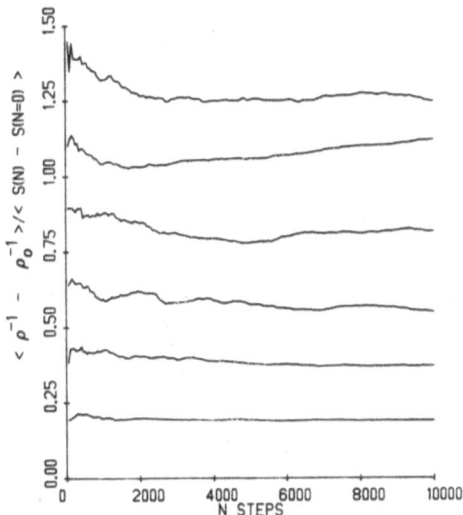

Fig. 6. Left hand side of Eq. (9) versus time for the square lattice with uniformly distributed site energies for (bottom to top): $T'' = 0.05$, 0.06, 0.07, 0.08, 0.09, and 0.1. Curves shifted (see Fig. 5).

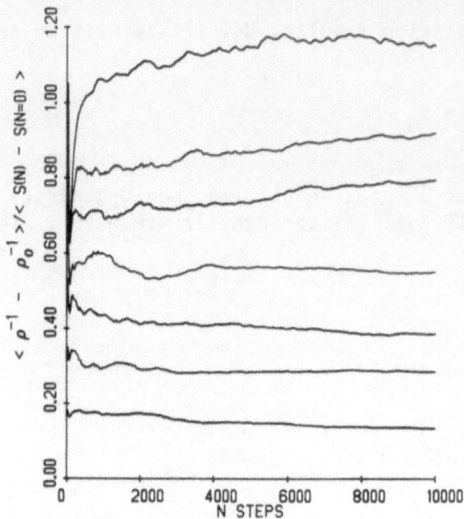

Fig. 7. Left hand side of Eq. (9) versus time for the simple cubic lat-
         tice with uniformly distributed site energies for (bottom to
         top): $T'' = 0.05, 0.055, 0.06, 0.065, 0.07, 0.1$ and for the
         isoenergetic simple cubic lattice. Curves shifted (see Fig. 5).

Eq. (9) is tested below for the planar Sierpinski gasket in Fig. (5) and
for the square lattice, Fig. (6), with uniformly distributed site energies; results from the simple cubic lattice are shown for both uniform,
Fig. (7), and Gaussian, Fig. (8), distributions. At early times deviations from a slope of zero are expected, but even at longer times few of
the curves are exactly horizontal. This may be due partially to statistical fluctuations in the simulation averages and partially to the fact that
Eq. (9) is a first order approximation.

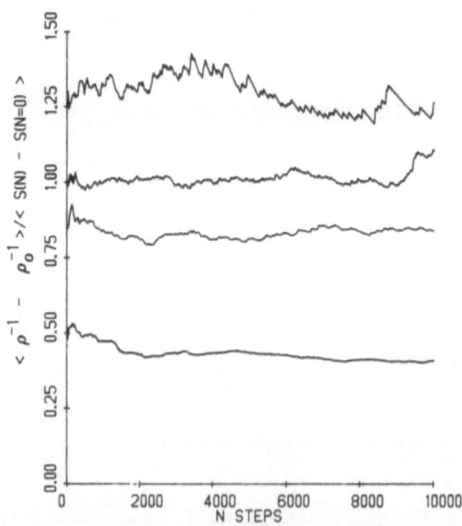

Fig. 8. Left hand side of Eq. (9) versus time for the simple cubic lat-
         tice with a Gaussian distribution of site energies for (bottom
         to top): $T'' = 0.05, 0.066, 0.083, 0.166$. Curves shifted (see
         Fig. 5).

CONCLUSIONS

As shown by the Sierpinski gasket simulations of single walker transport, the basic power law relations for the number of distinct sites visited by a random walker, Eq. (3) is a good first-order approximation even for energetically disordered systems with temperature-dependent Boltzmann weighting factors. The major difference between the exponential, Gaussian and uniform distributions of site energies is seen in the crossover behavior: it is most apparent for the uniform distribution and to a lesser degree for the Gaussian distribution. Simulations of the exponential distribution do not show a clear crossover, but it cannot be ruled out on the basis of these simulation results.

The stochastic approach (Smoluchowski-type) adopted here relates the macroscopic time-dependent rate coefficient to the microscopic transport of individual walkers, Eq. (7). It is a fairly good approximation even for energetically disordered systems. Eq. (5) provides a scaling approach for moving from single walker simulations to reacting walker simulations. This approach works well over a fairly broad time interval and is relatively insensitive to the detailed nature of the system considered. Furthermore, it emphasizes that the rate-determining factor in these diffusion-limited reactions is indeed due to transport constraints imposed by the medium on the individual walkers.

ACKNOWLEDGEMENTS

We would especially like to thank Dr. J. Klafter and Dr. S. E. Anacker for helpful discussions. Dr. S. E. Anacker's work on the computer graphics is greatly appreciated. This work was supported by NSF Grant No. DMR8303919.

REFERENCES

1. P. G. deGennes, C. R. Acad. Sci. (Paris) Ser. A 296 881 (1983).
2. P. Evesque, J. Phys., 44 1227 (1983).
3. P. Evesque and J. Duran, J. Chem. Phys. 80 3016 (1984).
4. P. W. Klymko and R. Kopelman, J. Lumin. a) 24/25 457 (1981), b) J. Phys. Chem. 86 3686 (1982), c) 87 4565 (1983).
5. L. W. Anacker, P. W. Klymko, R. Kopelman, J. Lumin 31+32 648 (1984).
6. R. Kopelman, P. W. Klymko, J. S. Newhouse, L. W. Anacker, Phys. Rev. B 29 3747 (1984).
7. L. W. Anacker and R. Kopelman, J. Chem. Phys. 81 6402 (1984).
8. L. W. Anacker, R. P. Parson, R. Kopelman, J. Phys. Chem. 89 4758 (1985).
9. L. W. Anacker, R. Kopelman, J. S. Newhouse, J. Stat. Phys. 36 591 (1984).
10. K. Kang and S. Redner, Phys. Rev. Lett. 52 955 (1984).
11. R. Kopelman, J. Hoshen, J. S. Newhouse, P. Argyrakis, J. Stat. Phys. 30 335 (1983).
12. G. Zumofen, A. Blumen, J. Klafter, J. Chem. Phys. 82 3198 (1985).
13. P. Meakin and H. E. Stanley, J. Phys. A17 L173 (1984).
14. D. C. Torney and H. M. McConnell, J. Phys. Chem. 87 1441 (1983).
15. D. C. Torney, J. Chem. Phys. 79 3606 (1983).
16. P. V. Elyutin, J. Phys. C, 17 1867 (1984)
17. S. Alexander and R. Orbach, J. Phys. Lett. (Paris) 43 1625 (1982).
18. R. Rammal and G. Toulouse, J. Phys. Lett. (Paris) 44 L13 (1983).
19. J. C. Angles d'Auriac, A. Benoit, and R. Rammal, J. Phys. A: Math. Gen. 16 4039 (1983).
20. R. Rammal, J. Stat. Phys. 36 547 (1984).
21. R. B. Pandy and D. Stauffer, Phys. Rev. Lett. 51 527 (1983).

22. P. Argyrakis, L. W. Anacker, R. Kopelman, J. Stat. Phys. _36_ 561 (1984).
23. J. Klafter, A. Blumen, G. Zumofen, J. Stat. Phys. _36_ 639 (1984).
24. M. F. Shlesinger, J. Stat. Phys. _36_ 639 (1984).
25. J. T. Bendler and M. Shlesinger, Macromolecules _18_ 591 (1984).
26. E. W. Montroll and G. H. Weiss, J Math. Phys. _6_ 167 (1965).
27. H. Scher and E. W. Montroll, Phys. Rev. B _12_ 2455 (1975).
28. H. Scher and M. Lax, Phys. Rev. B _7_ 4491, 4502 (1973).
29. J. Klafter and R. Silbey, Phys. Rev. Lett. _44_ 55 (1980).
30. G. Schönherr, H. Bässler, and M. Silver, Phil. Mag. B, _44_ 47 (1981).
31. P. B. Allen, J. Phys. C _13_ L667 (1980).
32. S. Alexander, J. Bernasconi, R. W. Schneider, R. Orbach, Rev. Mod. Phys. _53_ 175 (1981).
33. T. E. Orlowski and H. Scher, Phys. Rev. Lett. _54_ 220 (1985).
34. J. M. Marshall and A. C. Sharp, J. Non-Crys. Solids _35/36_ 99 (1980).
35. M. Silver, E. Snow, D. Adler, Solid State Communications _53_ 637 (1985).
36. V. M. Smoluchowski, Z. Physik. Chem. _92_ 129 (1917).
37. S. Chandrasekhar, Rev. Mod. Phys. _15_ 1 (1943).
38. P. Argyrakis and R. Kopelman, J. Chem. Phys. _72_ 3053 (1980).
39. H. B. Rosenstock, Phys. Rev. _187_ 1166 (1969), SIAM J. Appl. Math. _27_ 457 (1957).
40. R. P. Parson and R. Kopelman, Chem. Phys. Lett. _104_ 320 (1984).
41. B. B. Mandelbrot, The Fractal Geometry of Nature (W. H. Fr nan, San Francisco, 1983).

DYNAMICAL HOLE-BURNING REQUIREMENTS FOR FREQUENCY DOMAIN OPTICAL STORAGE

W. E. Moerner

IBM Almaden Research Center
K32/802D, 650 Harry Road
San Jose, CA 95120

INTRODUCTION

One potentially important application of laser spectroscopy and photochemistry of molecular and ionic defects in solids at low temperatures is the use of persistent spectral hole-burning to form a frequency domain optical storage system [1]- [5] . PHB (also called photochemical hole-burning although photophysical hole-burning also occurs) has the interesting property of allowing as many as 1000 or more bits of information to be stored in the volume irradiated by a single focused laser beam. PHB utilizes an additional dimension beyond x-y spatial dimensions to achieve this dramatic increase in areal storage density. In effect, various bits are addressed by the laser frequency or wavelength at which they are stored, hence the name "frequency domain optical storage" means the use of PHB for optical storage of digital data. Of course, the feasibility of such a data storage scheme depends critically upon having recording materials that undergo spectral hole-burning with certain well-defined dynamical characteristics. It is a stimulating and interdisciplinary challenge for the solid-state spectroscopist, photochemist, and laser physicist to find suitable mechanisms and to devise detection techniques that make this application possible.

The first requirement for spectral hole-burning is that the guest molecules (or other absorbing centers [6]) must be dispersed in a suitable transparent matrix and cooled to liquid helium temperatures. The molecules must be sufficiently rigid and sufficiently uncoupled from the host lattice that the lowest energy optical absorption (the "zero-phonon" absorption or the purely electronic transition) has appreciable oscillator strength. This requirement simply means that the Franck-Condon distortion in the electronic excited state cannot be too large. If the host matrix were a perfect crystal, the zero-phonon line would appear as in the upper half of Figure 1: all molecules would absorb at the same frequency with the same width, Γ_H, called the homogeneous width of the electronic origin. However, in a real crystal, glass, or polymer (see the lower half of Figure 1), slight differences in the environment around each molecule caused by local strains or nearby defects cause the various molecules absorb at slightly different wavelengths. For this case, the way in which the sample absorbs light in the lowest energy absorption can be viewed as a superposition of narrow absorption lines of width Γ_H from the various molecules distributed throughout the sample. The result is a broad, smooth absorption line of width Γ_I that is said to be <u>inhomogeneously</u> broadened.

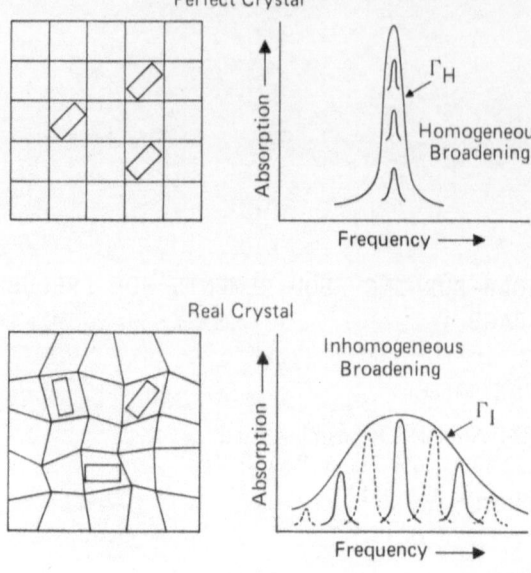

Fig. 1.   (Upper half) Schematic of absorbers dispersed in a perfect crystal. At low temperatures, the absorption line is homogeneously broadened with width Γ_H. (Lower half) Illustration of one source of inhomogeneous broadening in real solid matrices. The distribution of local environments leads to a distribution of center frequencies of absorption. The resulting lineshape has width Γ_I.

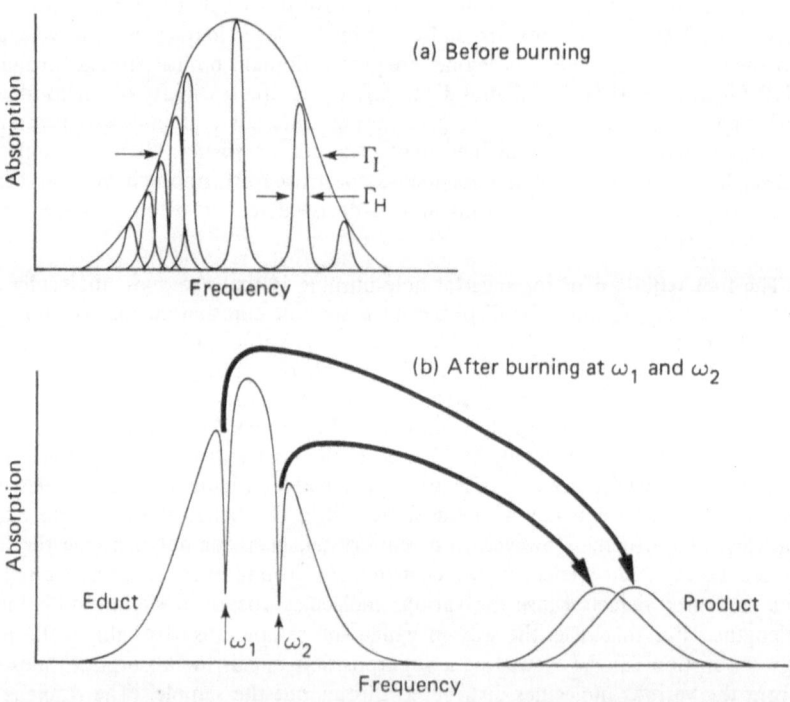

Fig. 2.   (a) Schematic of an inhomogeneously broadened absorption line at low temperatures. (b) Illustration of the formation of a spectral hole due to laser irradiation at ω_1 and ω_2.

THE BASIC PROCESS OF SPECTRAL HOLE FORMATION

If a laser beam with a well-defined frequency ω_1 irradiates such a sample, only those molecules that are in resonance with the laser can be excited (see Figure 2 (a) and (b)). If the laser linewidth is less than Γ_I or better yet less than Γ_H as is usually the case, one can selectively excite different classes of molecules simply by tuning the laser. The number of different classes of molecules that can be accessed in this way is on the order of the ratio of the inhomogeneous width to the homogeneous width, Γ_I/Γ_H, a factor which can range from 100 to 10^4. (It is precisely because we want this factor to be large that liquid helium temperatures are required. At higher temperatures, Γ_H becomes larger than Γ_I due to phonon scattering and other temperature-dependent dephasing effects.) If the laser is focused to a small spot, the size of that spot is limited by the wavelength of the light itself to be greater than roughly one micron in diameter. Even within the small volume irradiated by a one micron diameter laser beam, a narrowband tunable laser can easily probe different groups of molecules simply by changing the laser wavelength.

Now if the molecules undergo a photo-induced change when light is absorbed (see Figure 2(b)) such that the product does not absorb at the laser wavelength, the optical absorption at ω_1 decreases. Such light-induced changes may involve a photochemical change in the molecule itself where a product is formed [8] [9] or a light-induced change in the local environment near the molecule [10] [11] . The resulting decrease or "dip" at ω_1 in the absorption line is called a <u>spectral hole</u>. If the photoinduced change in the molecule is persistent on time scales of months or years at low temperatures, the spectral holes at various locations within the line can be used to encode binary information, where, for instance, the presence of a hole at a particular frequency within the inhomogeneous line might correspond to a binary "1" and the absence of a hole might correspond to a binary "0". Figure 3 gives an example of a 19 bit hole pattern written in the absorption line of free-base phthalocyanine molecules in a poly(methylmethacrylate) (PMMA) host [4]. Since in one laser spot many groups of molecules are available, 1000 or perhaps 10,000 bits can, in principle, be stored in the frequency domain in the volume illuminated by a focused laser beam, resulting in potential areal storage densities as large as 10^{11} bits/cm^2. This type of data storage system can feature fast random access and high data rates as well as high areal density, and is called a frequency domain optical storage system. As one can see, PHB utilizes many of the unique properties of laser radiation: narrow linewidth, spatial coherence, and tunability. The ability to use PHB in a future optical storage application depends heavily upon the dynamics of the photo-induced change that occurs in the absorbing centers such as hole width, quantum efficiency, presence of bottlenecks, etc.

GENERAL SYSTEMS AND MATERIALS ISSUES

It is to be emphasized that even though the frequency domain optical storage concept sounds simple, this idea can hardly be regarded as a fully practical technology at the present time. Indeed, no single material has been found that possesses all the required properties. In the rest of this paper, several crucial materials and engineering issues and their partial solutions will be described. Table 1 lists some of the properties that are required of the reading and writing system engineering and Table 2 list some of the required materials properties in order to produce practical data storage and retrieval. The middle column lists the material or technique with which the property has been demonstrated. For each entry in the table, the reader is urged to consult the references for more detail. In general, most of the engineering or systems requirements have been shown to be solvable within the current state of the art, although single-mode diode lasers with wider tuning ranges would be desirable. In particular, recent research on FM spectroscopy has demonstrated the shot-noise-limited sensitivity of this method for detection of unmodulated absorptions [21].

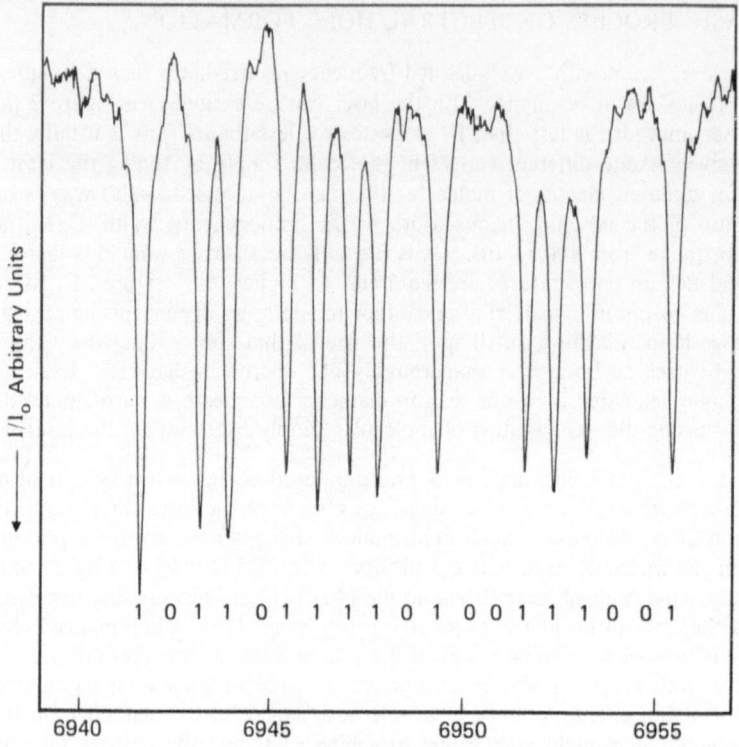

1 0 1 1 0 1 1 1 1 0 1 0 0 1 1 1 0 0 1

Wavelength in Angstroms

Fig. 3.   Portion of the absorption spectrum for free-base phthalocyanine molecules in P.M.M.A showing a sequence of holes burned in the absorption line (after Reference [4], by permission).

Table 1. Engineering Requirements

Property	Material or Technique	Reference
Hole detection	FM spectroscopy, λ modulation, HUMPH, FREMPOLSPECT	[12] [13] [14] [15]
Reading and writing with diode lasers	current and temperature tuning	[16] [17]
Focus/servo in liquid He	Ronchi grating, dither	[18]
No crosstalk between adjacent spots	measured: R' in LiF	[19] [20]

Any reasonable system design also places demands on the dynamical properties of the photoactive hole-burning material itself, and Table 2 summarizes some of the principal requirements. Most of these requirements involve fairly obvious considerations. For instance, the material must undergo photochemistry at GaAlAs laser wavelengths, because these lasers are tunable, inexpensive, compact, and readily available commercially. To give another example, holes must be burned in times on the order of 30 ns in order to handle the high data rates expected for a large data base. Recently, detectable holes have been burned in less than 100 nsec using a high sensitivity organic system and FM spectroscopy detection [27]. The existence of most of the required properties in the upper half of the table has been separately demonstrated in a number of inorganic and organic systems, which attests to the high rate of recent progress in this area.

SINGLE-PHOTON PROCESSES

Let us now focus on the entry in Table 2 that is third from the bottom: a material in order to be useful in a practical storage system must simultaneously show all the required properties: the ability to form deep holes in short burning times and yet allow fast reading at high signal-to-noise ratios with focused beams. This requirement places several well-defined constraints on the dynamical properties of the hole-burning mechanism. For example, a material with low hole-burning efficiency would be quite easy to read without serious destruction of the written holes, but such a system would be difficult to write with high contrast in short burning times. Conversely, a system that shows fast burning due to a high quantum efficiency for hole production would be difficult to read without burning of the unwritten centers by the tightly focused reading beam.

To understand this problem more fully, a thorough analysis of the coupled reading-writing problem in small spots for materials with monophotonic of single-photon hole-burning mechanisms has been recently completed [28]. Figure 4 schematically shows the energy level structure appropriate to single-photon mechanisms. An incident photon flux F is absorbed with an absorption cross section σ. A given center that has absorbed a photon can either decay to the ground state with rate Γ_1 or undergo the transformation leading to hole-burning with probability or quantum efficiency η [29]. This simple level diagram contains all the salient features of monophotonic photochemical processes. The essential problem with single-photon processes is that there is no threshold in the hole formation mechanism. The process of hole detection requires the absorption of photons by the remaining unburned centers, and if high powers are necessary to detect the dip in the absorption line with adequate signal-to-noise, the hole pattern will be destroyed by the reading laser beam (i.e., a "trench" will be formed over the spectral region probed by the reading laser). This optimization problem has been analyzed in detail to determine whether any combination of single-photon materials parameters would yield acceptable reading performance [28].

Figure 5 shows the results of this dynamical analysis of single-photon mechanisms. The two fundamental materials parameters are naturally the hole-burning quantum efficiency η and the low temperature peak absorption cross section σ. A material with specific values of η and σ would be represented by a point on the η-σ plane shown in the figure. Furthermore, since the analysis requires that the absorption coefficient α_0 of the sample be fixed at a constant value that optimizes the signal-to-noise ratio, the concentration of centers necessary to keep α_0 fixed (top axis) must decrease as the cross section increases. In other words, for given η and σ, the concentration listed on the top axis must be achieved in order to optimize the signal-to-noise ratio. The lower right triangular region represents the class of materials parameters that would not yield acceptable signal-to-noise ratios for any number of reads. In other words, a useful single-photon material must have low absorption cross section and high quantum efficiency, with sufficient solubility in the host to yield the concentrations shown on the

Table 2. Material Requirements

Property	Material or Technique	Reference
Δv_{hole} = 100-500 MHz	color centers, H_2Pc-PE,...	[7] [22]
GaAlAs compatibility	R' in LiF(8330 Å) H_2Pc-PE in H_2SO_4	[23] [24]
Reversible burning	H_2Pc, H_2P, ...	[9] [25]
Long hole lifetime at low temperatures	Quinizarin in glasses	[26]
Fast burning ($\simeq$30 ns/bit)	H_2Pc-PE	[27]
Fast burning, high SNR, fast reading, focused spot	difficult for single-photon mechanisms	[28]
Gated hole-burning	Sm^{2+}-BaClF, carbazole in boric acid	[31] [32]
Room temperature cycling	Sm^{2+}-BaClF	[31]

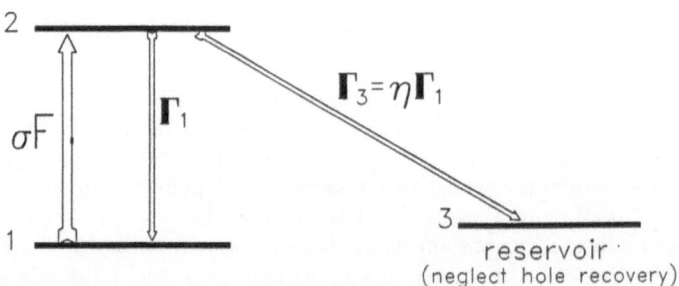

Fig. 4.   Schematic energy level diagram for absorbing centers with a single-photon hole-burning mechanism. Level 1 is the ground state, level 2 is the excited state, and level 3 is a permanent reservoir ground state that schematically depicts the hole formation process. The various pumping and decay rates are defined in the text.

upper axis of the figure. Within the upper left allowed region, the lines represent contours of constant numbers of reads. The achievable number of reads varies from 1 to greater than 10^4; again superior performance results from low cross sections and high quantum efficiencies. (η values greater than 0.1 are not allowed, because hole broadening would occur due to excessive photochemistry, and σ values less than 10^{-15} cm^2 would require prohibitively high densities of centers.)

The results in Figure 5 show that a new challenge exists for scientists in the field of PHB: find single-photon materials that have values of quantum efficiency and cross section that fall within the allowed region, as well as solubilities that allow concentrations shown along the top axis. One might look for hole-burning in partially allowed transitions, such as n-π^* transitions of organic molecules and d-f transitions of divalent rare earth systems. However, the parameter space of useful η-σ values is somewhat small.

GATED PHB

One way around this problem would be to consider those materials that do not suffer from the intrinsic limitations of single-photon, monophotonic processes with no threshold. One such class of materials are those with two-step PHB mechanisms, called gated mechanisms in the second to the last entry in Table 2. Figure 6 shows the general idea of gating. The wavelength λ_1 excites a homogeneous packet within an inhomogeneously broadened line. If no external field is present, the center returns to the original ground state without forming a spectral hole. However, in the presence of λ_1 and some external field, the center undergoes the transformation leading to hole-burning. This is the origin of the term "gating": the external field acts as a gate on the hole-burning process. The hole may be detected using λ_1 alone, since hole detection is merely probing the ground state distribution of those centers that did not react to form the spectral hole. Since the external field is not present during the reading process, hole detection may then be nondestructive. In effect, gated mechanisms add a "threshold" to the writing process, which uncouples the reading and writing processes. The external field may be a second photon of a different wavelength (photon-gating [30]) or the gating could perhaps be achieved by any other external field, such as electric field, magnetic field, stress field, and the like.

Recent materials research at IBM has been devoted to a search for gated or two-color or photon-gated PHB mechanisms in inorganic as well as organic materials. In photon-gated PHB, two photons (of different wavelengths) are required for the photoinduced change leading to hole-burning (writing). Last year, photon gated PHB was observed for Sm^{+2} ions in BaClF crystals [31]. The mechanism is thought to involve a three-level process similar to that shown schematically in Figure 7. The first wavelength, λ_1, excites the system from the ground state to an intermediate level. Extended irradiation at λ_1 produces essentially no hole production, but brief periods of simultaneous irradiation with λ_2 allows deep holes to be burned at λ_1. The second photon is thought to excite the ion from level 2 to the conduction band or to an autoionizing level and the photoejected electron is subsequently trapped in the host matrix.

This material has a further exciting and unexpected property: a pattern of holes burned at low temperature persists even after cycling up to room temperature and back down to helium temperatures. Apparently, the electrons are trapped at sites with very high barriers, and the relaxation of strains upon cycling is small enough to prevent loss of the site selection. This discovery shows that materials exist that can relieve one of the most serious concerns with frequency domain optical storage: volatility of the stored data. Further experiments are in progress to understand this novel process, and the reader is urged to consult the references for more detail [31].

Moreover, two-color photon gated PHB has also been observed in an organic material composed of carbazole molecules in boric acid glasses [32]. This important

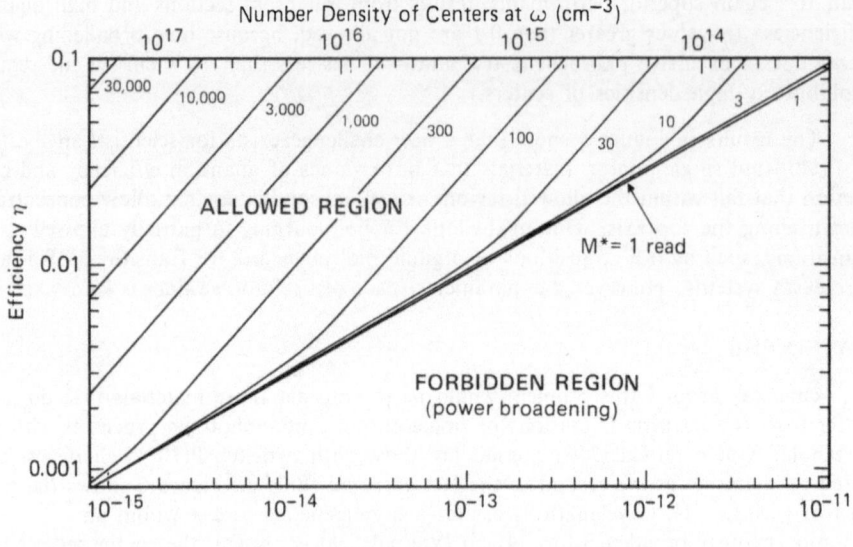

Fig. 5. Allowed region of efficiency η, cross section σ, and number density of centers at ω, in order for the first read to yield acceptable signal-to-noise ratio. M^* is the number of reads achievable for a given material. The contours in the upper left portion of the figure represent contours of constant M^*.

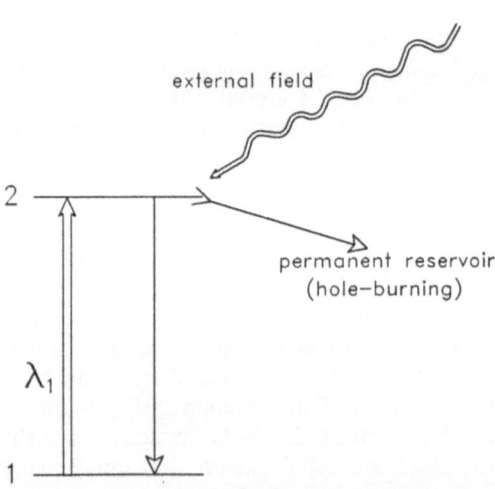

Fig. 6. Illustration of basic scheme for gated, two-step PHB.

result proves that gated mechanisms exist in organics as well as inorganics. In addition, one drawback of the inorganic material is the low absorption cross section, which prevents the achievement of usable optical densities in thin film samples. This problem is absent in the organic system, although the organic material requires near ultraviolet wavelengths for site-selection. The presumed mechanism is depicted schematically in Figure 8. Upon excitation in the singlet-singlet origin with $\lambda_1 = 335$ nm, the molecule undergoes intersystem crossing with a high yield to form triplets. From level 3 (the lowest triplet, T_1), the molecules return to the original ground state if λ_2 is not present, and no hole is formed. However, in the presence of $\lambda_2 = 360-405$ nm, holes are formed at the singlet excitation wavelength, λ_1. Excitation spectra of λ_2 and the product absorption spectrum suggest that the mechanism is photoionization of the molecule and trapping of the ejected electron in the boric acid glass matrix. The special property of boric acid glass in reducing the ionization potential of the guest is critically important for the two-step photoionization process; no two-color holes are formed if the host is changed to poly(methyl methacrylate), for example.

AREAS FOR FUTURE RESEARCH

These two new examples of gated spectral hole-burning have opened up a new class of materials for PHB, and considering the limitations on single-photon materials, the search for gated mechanisms should be an important area for future studies. Of course, the previous observations of gated PHB can only be regarded as proofs-of-principle, because the materials showing gating do not possess all the other required properties for frequency domain optical storage. Gated PHB should be observable in a variety of other systems that may offer improved properties. The search must now concentrate on finding the largest number of useful properties in one material active at diode laser wavelengths. Considering that 8-10 years of basic research was necessary on single-photon processes in order to attain the present level of understanding, gated processes deserve an equal amount of effort and attention. An generalized analysis of the reading and writing performance of gated PHB materials has recently been performed [33] that will aid in the continuing search for optimal mechanisms.

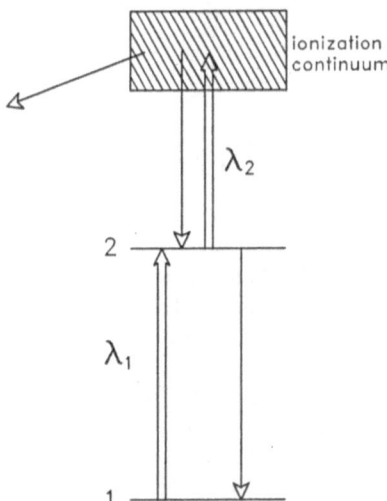

Fig. 7.   Illustration of gated PHB for a three-level system.

49

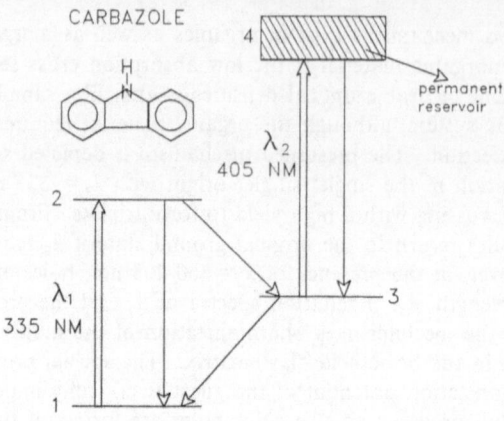

Fig. 8. Illustration of the four-level gated PHB mechanism for carbazole molecules in boric acid glass. Level 1 is the singlet ground state S_0, level 2 is the singlet excited state S_1, level 3 is the lowest triplet state T_1, and the levels labeled 4 represent conduction band or autoionizing states that lead to photoionization.

Nevertheless, single photon processes with η and σ values in the upper left corner of Figure 5 would satisfy many of the requirements for frequency domain optical storage applications. Since years of experience has already been gained in single-photon materials, and considering that gated PHB materials are extremely new, a parallel effort to find optimal single-photon processes would also be useful. For any material, the issue of erasability must be considered to see if a sufficient number of erase cycles can be achieved. Furthermore, the possible existence of deleterious heating effects must be ascertained to assure that reading and writing power levels are not so high that stored holes are broadened or erased. The future feasibility of a frequency domain optical storage system rests on the discovery of single-photon materials with the required values of cross section and quantum efficiency, or upon optimizing gated mechanisms for spectral hole-burning.

Acknowledgment

This work was supported in part by the Office of Naval Research.

REFERENCES

1. G. Castro, D. Haarer, R. M. Macfarlane, and H. P. Trommsdorff, "Frequency selective optical data storage system," U. S. Patent No. 4,101,976, (1978).
2. G. C. Bjorklund, W. Lenth, and C. Ortiz, Proc. Soc. Photo-Opt. Instr. Eng. 298, 107 (1981).
3. D. Haarer, Proc. Soc. Photo-Opt. Inst. Engr. 177, 97 (1979).
4. A. R. Gutierrez, J. Friedrich, D. Haarer, and H. Wolfrum, IBM J. Res. Devel. 26, 198 (1982), and references therein.
5. W. E. Moerner, J. Molec. Elec., 1, 55 (1985).

6. In this article, we will refer to the absorbing centers as "molecules", even though the centers could be color centers [7] or other types of defects.

7. R. M. Macfarlane, R. T. Harley, and R. M. Shelby, Rad. Effects 72, 1 (1983), and references therein.

8. J. Friedrich and D. Haarer, Angew. Chemie 23, 113 (1984), and references therein.

9. L. A. Rebane, A. A. Gorokhovskii, and J. V. Kikas, Appl. Phys. B29, 235-250 (1982).

10. G. J. Small, in Spectroscopy and Excitation Dynamics of Condensed Molecular Systems, V. M. Agranovitch and R. M. Hochstrasser, editors, (North-Holland, Amsterdam, 1983), pp. 515-554.

11. W. E. Moerner, A. J. Sievers, R. H. Silsbee, A. R. Chraplyvy, and D. K. Lambert, Phys. Rev. Lett. 49, 398 (1982); W. E. Moerner, A. R. Chraplyvy, A. J. Sievers, and R. H. Silsbee, Phys. Rev. B 28, 7244 (1983).

12. G. C. Bjorklund, M. D. Levenson, W. Lenth, and C. Ortiz, Appl. Phys. B32, 145 (1983).

13. P. Pokrowsky, W. Zapka, F. Chu, and G. C. Bjorklund, Opt. Commun. 44, 175 (1983).

14. A. L. Huston and W. E. Moerner, J. Opt. Soc. Am. B: Opt. Phys. 1, 349 (1984).

15. M. Romagnoli, M. D. Levenson, and G. C. Bjorklund, Opt. Lett. 8, 635 (1983); M. Romagnoli, M. D. Levenson, and G. C. Bjorklund, J. Opt. Soc. Am. B: Opt. Phys. 1, 571 (1984).

16. P. Pokrowsky, W. E. Moerner, F. Chu, and G. C. Bjorklund, Opt. Lett. 8, 280 (1983).

17. P. Pokrowsky, W. E. Moerner, F.Chu, and G. C. Bjorklund, Proc. Soc. Photo-Opt. Instr. Engr. 382, 202 (1983).

18. D. J. Bernays, Proc. SPIE, Vol. 498, 175 (1984).

19. F. M. Schellenberg, W. E. Moerner, M. D. Levenson, G. C. Bjorklund, and D. J. Bernays (1984), Conference on Lasers and Electro-optics Technical Digest, June 19-22, 1984, Anaheim, California, paper ThI41.

20. B. H. Schechtman, G. C. Bjorklund, and W. E. Moerner, IBM Research Report # RJ4128, December 8, 1983.

21. M. Gehrtz, W. E. Moerner, and G. C. Bjorklund, IBM Research Report # RJ 4574, January 21, 1985.

22. W. E. Moerner, in Proceedings of the International Conference: Lasers '83, R. C. Powell, editor, (STS Press, McLean, VA, 1983), p. 489.

23. W. E. Moerner, F. M. Schellenberg, and G. C. Bjorklund, Appl. Phys. B28, 263 (1982); W. E. Moerner, P. Pokrowsky, F. M. Schellenberg, and G. C. Bjorklund (submitted to Phys. Rev. B).

24. H. W. H. Lee, A. L. Huston, M. Gehrtz, and W. E. Moerner, Chem. Phys. Lett. 114, 491 (1985).

25. S. Völker and J. H. van der Waals, Molec. Phys. 32, 1703 (1976); S. Völker and R. M. Macfarlane, IBM J. Res. Devel. 23, 547 (1979).

26. W. Breinl, J. Friedrich, and D. Haarer, Chem. Phys. Lett. 106, 487 (1984).

27. M. Romagnoli, W. E. Moerner, F. M. Schellenberg, M. D. Levenson, and G. C. Bjorklund, J. Opt. Soc. Am. B: Optical Physics 1, 341 (1984).

28. W. E. Moerner and M. D. Levenson, J. Opt. Soc. Amer. B: Optical Physics 2, 915 (1985).

29. W. E. Moerner, M. Gehrtz, and A. L. Huston, J. Phys. Chem. 88, 6459 (1984).

30. D. M. Burland, F. Carmona, G. Castro, D. Haarer, and R. M. Macfarlane, IBM Tech. Discl. Bull. 21, 3770 (1979).

31. A. Winnacker, R. M. Shelby, and R. M. Macfarlane, Opt. Lett. 10, 350 (1985).

32. H. W. H. Lee, M. Gehrtz, E. Marinero, and W. E. Moerner, Chem. Phys. Lett. 118, 611 (1985).

33. W. Lenth and W. E. Moerner, Opt. Commun. 58, 249 (1986).

ZERO-PHONON LINES, PHOTOBURNING OF SPECTRAL HOLES,

INFORMATION STORAGE, PHOTON ECHO, IN SOLID SYSTEMS

Karl K. Rebane

Institute of Physics
Academy of Sciences of the Estonian SSR
202400 Tartu, Riia St. 142, USSR

1. The photoburning of spectral holes (PBSH) [1,2] serves not only as an effective tool for high-resolution high-sensitivity spectroscopy and photochemistry but also opens possibilities for novel applications, such as narrow-band spectral filters and data processing [3,4,5].

The foundation stone of PBSH is the very sharp and intense zero-phonon line (ZPL) in optical spectra of impurity centers in solid matrices [6].

The theory shows that the ZPL of the purely electronic transition (PEL) of a single molecule (atom, ion) as an impurity center or of a body of absolutely identical centers – the homogeneous ZPL – actually is an optical analog of the very narrow and sensitive Mössbauer γ-resonance ZPL. Both ZPLs are free of Doppler broadening and the decay-time τ_{opt} determined linewidths δ_o should be available in (stationary) experiments.

In the case of optical ZPLs the tremendous inhomogeneous broadening (overwhelming δ_o by a factor of $10^3 - 10^4$ or more) has to be eliminated. This is being done by methods of laser spectroscopy, PBSH being the most effective in several respects.

2. After the inhomogeneous broadening is eliminated, the optical purely electronic ZPL really becomes the analog of the Mössbauer line. The ZPL widths $\delta_o(T) \sim \tau_{opt}^{-1}$ are determined by the lifetime τ_{opt} ($\tau_{opt}^{-1} = \tau_{decay}^{-1} + (2\tau)_{phase\ memory}^{-1}$, the $\tau_{phase\ memory}$ decreasing rapidly with temperature) of the excited electronic state of the impurity

and at 1.8-4K the $\delta_o(T)$ lie within $10^{-3} - 10^{-4}$ cm^{-1} for allowed purely electronic transitions.

On the other hand, PBSH turns the inhomogeneously broadened impurity absorption band into a useful feature of the optical properties of the system impurity + matrix.

3. Given the unique properties of the ZPL media (inhomogeneous spectral bandwidth 100-1000 cm^{-1}; linewidth of PEL 10^{-3}-10^{-4}cm^{-1}, and even less for forbidden transitions), the transition frequency to linwidth relation (about $10^7 - 10^9$) can be utilized via PBSH in two complementary ways.

3.1. It is possible to burn in an IHB band up to 1000-10,000 holes at different frequencies using narrow lines of continuous-wave lasers and thus create narrow-band optical filters of a very special kind - frequency-selective optical memories of high capacity. Some theoretical and practical limitations of these possible memories are discussed.

3.2. Excitation of the PEL medium by a picosecond pulse creates a broad (2-5 cm^{-1}) hole whose frequency distribution represents the intensities of the harmonics in the pulse. The information about the phases of the Fourier components of the pulse spectrum is lost. In the case of two successive pulses (the phases of one of them know, e.g., a delta pulse) separated by a time interval shorter than the phase relaxation time $\tau_{\text{phase memory}}$ of the excited electronic state of the impurity system, the phase relations between the pulses will be fixed as well by the burnt-in picture of the spectral holes. Very-high-relative-intensity (up to 50%) coherent optical free-decay signals can be stimulated by weak picosecond excitation of these persistent spectral holograms.

4. Doppler scanning as a method to study narrow holes is developed and applied to fast spectral domain PBSH investigations. The presence of slow diffusion processes (most probably activated by illumination) but still fast in comparison with the hole burning + hole measurement time τ_{HB} has been demonstrated: the hole width decreases with decreasing τ_{HB}.

5. Temperature dependence of the hole widths $\delta_H(T)$ burnt in by the $\lambda_{\text{exc}} = 620.1$ nm laser line of the inhomogeneous body of H_2-octaethylporphin impurity molecules in polysterene glass at 0.05-1.5K was investigated

[7]. At 0.05K $\delta_H(T)$ = 26 MHz $(0.86 \cdot 10^{-3} \text{cm}^{-1})$, still considerably exceeding the limit value $\delta_H(0) = 2\delta_o(0)$ = 10.6 MHz $(0.35 \cdot 10^{-3} \text{ cm}^{-1})$ determined by the measured decay time of the S_1 state τ_{decay} = 29.6 nsec.

$\delta_H(T)$ increases at T < 0.1K as $T^{1.75 \pm 0.24}$, at T > 0.2K as $T^{1.18 \pm 0.06}$, and in the temperature interval 0.1-0.2K approximately as $T^{0.6 \pm 0.1}$. It should be pointed out that the experimental curve seems to have two crossover points: at 0.1K and 0.2K. Extrapolation of the dependence $T^{1.75}$ to lower temperatures shows that the $\delta_H(0)$ limit will be reached somewhat below T = 0.01K.

This temperature dependence (two crossover points) indicates at least two different mechanisms of the temperature broadening of the homogeneous widths of the ZPL in the system under consideration. At temperatures T > 0.1K the main reason of the broadening might be "fast" (in comparison with τ_{decay}) interaction with low-frequency excitations - phonon-like lattice excitations whose density is strongly amplified by two level system splittings. This mechanism can create a "real homogeneous" broadening. At T > 0.2K the "slow" mechanism of spectral diffusion caused by rearrangements (slow in comparison with τ_{decay} but fast in comparison with the hole burning + hole detection time) and resulting in "measurement-induced inhomogeneous" broadening becomes overwhelming [8].

6. Spectrally selective depopulation of nonselectively populated S_1-levels in photoactive impurities of octaethylporphin in polystyrene at 5K by stimulated emission is demonstrated. The stimulated emission restricts the BH at the laser frequency, while at other frequencies the broad band excitation works rather efficiently. It results in formation of stable antiholes $(\delta_H \sim 3 \text{ cm}^{-1}$, laser linewidth $\delta_L \sim 1 \text{ cm}^{-1})$ in absorption and fluorescence spectra and indicates the inactivity of the ground state vibrations in the impurity phototransformation. The method may serve as a tool to study the homogeneous widths of vibronic fluorescence lines [9].

7. PBSH by a series of pairs of picosecond pulses separated in time by less than the phase relaxation time $\tau_{phase\ memory}$ is shown to result in permanent spectral gratings in the sample transparency spectrum [10].

Coherent response of such spectral gratings to resonant picosecond excitation contains series of delayed picosecond pulses - photochemically accumulated stimulated photon echo (PASPE) [11,12].

Principles of holographic time-and-space domain picosecond optical data storage based on PASPE as well as picosecond wave conjugation are developed and demonstrated experimentally [13-15].

REFERENCES

1. A.A. Gorokhovskii, R.K. Kaarli, and L.A. Rebane, Pisma JETP, 20, 474, 1974.
2. B.M. Kharlamov, R.I. Personov, and L.A. Bykovskaya, Optics Commun., 12, 191, 1974.
3. L.A. Rebane, A.A. Gorokhovskii, and J.V. Kikas, Applied Physics, B29, 235, 1982.
4. J. Friedrich and D. Haarer, Angewandte Chemie (Internatl. Ed.) 23, 113, 1984.
5. W.E. Moerner, ed., Persistent Spectral Hole Burning, Science and Applications, Springer-Verlag, Berlin, 1988; O. Sild and K. Haller, Zero-Phonon Lines and Spectral Hole Burning in Spectroscopy and Photochemistry, Springer-Verlag, Berlin, 1988.
6. K.K. Rebane, Impurity Spectra of Solids, Plenum Press, New York, 1970.
7. A.A. Gorokhovskii, V.H. Korrovits, V.V. Palm, and M.A. Trummal, JETP Lett. 42, 307, 1985.
8. K.K. Rebane, J. Luminescence, 31-32, 744, 1984.
9. J. Kikas and I. Sildos, Chem. Phys. Lett., 114, 44, 1985.
10. A.K. Rebane, R.K. Kaarli, and P.M. Saari, Optika i Spektr., 55, No. 3, 405, 1983.
11. A.K. Rebane, R.K. Kaarli, and P.M. Saari, JETP Lett., 38, 383, 1983.
12. A. Rebane, R. Kaarli, P. Saari, Anijalg, and K. Tipmann, Optics Commun., 47, No 3, 173, 1983.
13. P.M. Saari, R.K. Kaarli, and A.K. Rebane, Sov. F. Quantum Electron, 15, 443, 1985.
14. P.M. Saari and A.K. Rebane, Izv. AN Est. SSR, Fiz. Matem., 33 No. 3, 322, 1984, (in Russian).
15. P. Saari, R. Kaarli, and A. Rebane, F. Opt. Soc. Am., B33, 527, 1986.

EVOLUTION OF SPECTRAL HOLES ON

LOGARITHMIC TIME SCALES

J. Friedrich and D. Haarer

Physikalisches Institut
Universität Bayreuth
D-8580 Bayreuth, FRG

Persistent spectral holes are extremely sensitve probes for long time tails in the relaxation patterns of amorphous solids. We succeeded recently in measuring the time evolution of photochemical and photo-physical holes in alcohol glasses at temperatures around 1K /1,2/. The photoactive probe molecules were 1,4 dihydroxyanthraquinone (DAQ) and tetracene (T). In alcohol glass, DAQ undergoes a light induced proton transfer reaction, whereas the detailed nature of the phototransformation of T in alcohol glass is not known (for a recent review see /3/). We measured the area of the holes, their width, their shape and their degree of polarization /4/. The experimental observation period was 1 week. The results can be summarized as follows:

1) the area A(t) <u>decreases</u> linearly on a logarithmic time scale (Fig. 1)
2) the width <u>increases</u> linearly on a logarithmic time scale (Fig. 3)
3) in deuterated glass, the relaxation processes occur on a time scale which is orders of magnitudes slower (Fig. 1 and Fig. 3)
4) the degree of polarizsation remains constant (Fig. 1 and Fig. 2)
5) the photochemical (DAQ) and the photophysical (T)-system show similar relaxation patterns (Fig. 1 and Fig. 2).

The above results can be interpreted very consistently within the frame of the TLS-model of the amorphous state /5/:

1) <u>The relaxation of the hole area</u>: From low temperature specific heat experiments it is well known that the energies of two level tunnelling states (TLS) are uniformly distributed (for a review see /6/). Consistent with a uniform distribution in energy is a uniform distribution

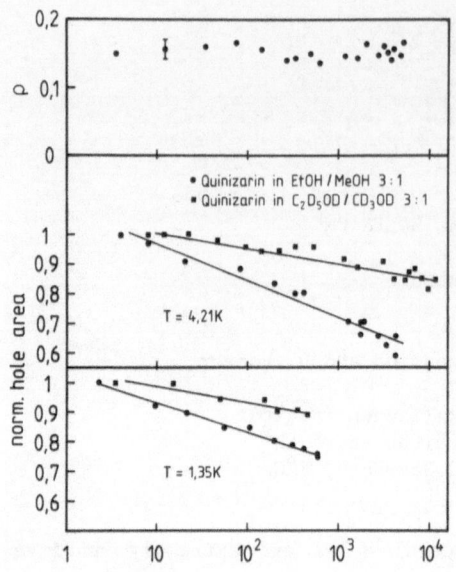

Fig. 1. Area and degree of polarization of a photochemical hole as a function of time. System: Quinizarin in alcohol glass.

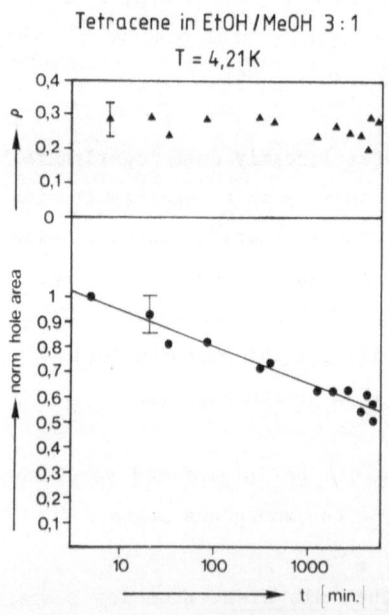

Fig. 2. Area and degree of polarization of a spectral hole as a function of time. System: Tetracene in alcohol glass.

in both parameters, which characterize a TLS, namely the energy asymmetry Δ and the tunnel parameter λ . A consequence of a uniform distribution of Δ and λ is a very asymmetric distribution of tunnelling relaxation rates R, roughly given by /7/

$$P\ (R) \sim R^{-1} \tag{1}$$

Equ. 1 is the starting point of our theoretical model /1,2/. Since the above distribution has to be normalized, there exist two cut-off values R_{min} and R_{max}, representing the slowest and the fastest tunnelling processes of the system considered. Once R_{min} and R_{max} are known the complete dynamics of the TLS-system under consideration is known. For the first time, we could estimate these parameters from hole burning experiments. Our arguments are the following: In both cases, quinizarin and tetracene, the photoproduct is trapped in some tunnelling state which is subject to a distribution of relaxation rates according to (1), reflecting the local disorder. At time t one observes mainly those centers relax which are governed by rates on the order $R = 1/t$. Hence, the hole area is given by integrating the above rate distribution from R_{min} to $R = 1/t$. This integral yields the number of tunnelling centers being, at time t, in the photoproduct state, and, thus, is proportional to the hole area. One derives for the hole area A(t) a logarithmic decay law:

$$A(t)/A_1 = 1 - \left[\ln R_1/R_{min}\right]^{-1}\ln t/t_1 \tag{2}$$

A_1 normalizes the decay function at the time t_1. The slope of the above linear relation is given by the logarithm of the two rates R_1/R_{min}. $R_1 = 1/t_1$ is an experimental parameter on the order of 1/min. Hence, R_{min}, can be directly determined from the measured slope. The dispersion of rates R_1/R_{min} is huge. We get for quinizarin in the protonated and in the deuterated glass, respectively, the values $[R_1/R_{min}]_H = 10^8$, $[R_1/R_{min}]_D = 10^{19}$. For tetracene in ethanol we found: $R_1/R_{min} = 10^{11}$.

2) <u>The time evolution of the linewidth</u>: The time evolution of the linewidth of the DAQ-system occurs on the same time scale as the decay of the hole area. To explain our experimental data we employ the concept of spectral diffusion. We consider a molecule, in the educt state, which has, at a time t_1, a sharp transition frequency. As time evolves, the molecules relaxing from the product to the educt state create strain fields which lead to a diffusion of excitation energies of the probe molecules in frequency space. As shown by several authors /7,8/, the

lineshape of the diffusing molecules is, under certain conditions, Lorentzian. Its width is proportional to the number $n(t-t_1)$ of molecules having 'flipped' within the time interval $t-t_1$:

$$\gamma_D \ (t-t_1) \sim n \ (t-t_1) \qquad\qquad (3)$$

If we consider the case $t_1 = 1/R_{max}$, then the above equation represents the full width due to spectral diffusion. Very similar to the procedure outlined above $n(t-1/R_{max})$ can be determined by integrating equ. 1 from $R = 1/t$ to $R = R_{max}$. The result is again a logarithmic law (Fig. 3) with a slope factor which depends on several experimental parameters like the temperature T, the total number of photochemically converted centers N and the concentration C.

$$\gamma_D = f(T,N,C) \left[\ln R_{max}/R_{min} \right]^{-1} \ln R_{max} t \qquad (4)$$

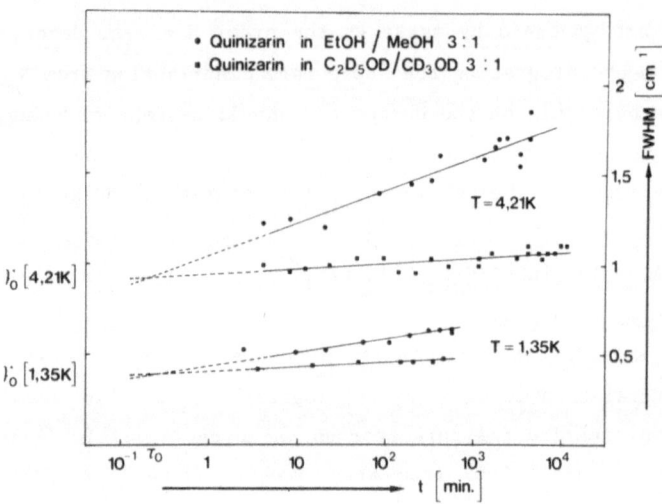

Fig. 3. Hole width as a function of time. System: Quinizarin in alcohol glass.

The measured width γ is the sum of a diffusional width and a time independent contribution γ_0 (which may be the homogeneous width). If one succeeds in measuring separately the diffusional width $\gamma_D = \gamma - \gamma_0$, then it is easy to show /2/ that a reduced plot $\gamma_D (t)/\gamma_D(t_1)$ yields a logarithmic law with a slope factor determined solely by R_{max}, the maximum rate constant:

$$\gamma_D(t)/\gamma_D(t_1) = (\gamma(t)-\gamma_0)/(\gamma(t_1)-\gamma_0) = 1+[\ln R_{max}/R_1]^{-1}\ln t/t_1 \qquad (5)$$

R_{max} can be determined experimentally from the measured slope. In our experiments we estimated γ_0 from the deuteration effect. As seen in Fig. 3, the extrapolated dependencies of the protonated and deuterated sample show a cross over at a time τ_0 of 12 s. Clearly τ_0 marks the onset of spectral diffusion and thus $\gamma(\tau_0) = \gamma_0$. Equ. 5 allows to give an estimate of the maximum rate constant. For quinizarin in the EtOH/MeOH glass we get $R_{max} \approx 0.025$ s^{-1}. We stress that R_{max} does not show any significant dependence on deuteration.

From the anisotropy and the isotope effect we get some information on the nature of the TLSs involved. The constant anisotropy as a function of time tells us that the dye molecules in the educt state are not directly involved in the TLS dynamics. The isotope effect in DAQ shows that the TLS dynamics is due to proton tunnelling. It further allows to make an estimate of the maximum barrier heights. These are much higer than the thermal energy at the glass transition temperature, indicating that barrier formation is most likely a collective phenomenon.

SUMMARY AND CONCLUSION

We have shown that the tunnelling relaxation processes of low temperature glasses, which lead to time dependent features of spectral holes, are well described by logarithmic laws which reflect a wide distribution of relaxation rates.

The logarithmic laws hold, of course, only within a certain time range. However, in case the dispersion of rates is large, this time range covers many orders of magnitude /9/. The dispersion of rates can be directly determined from the slope of the logarithmic decay law.

ACKNOWLEDGEMENT

The authors acknowledge a grant from the Stiftung Volkswagenwerk.

References

/1/   W. Breinl, J. Friedrich, D. Haarer, Chem.Phys.Lett. 106,487 (1984)
/2/   W. Breinl, J. Friedrich, D. Haarer, J.Chem.Phys. 81,3915 (1984)
/3/   J. Friedrich, D. Haarer, Ang.Chem.Int.Ed.Engl. 23,113 (1984)
/4/   W. Köhler, W. Breinl, J. Friedrich, J.Chem.Phys. 82,2935 (1985)
/5/   P.W. Anderson, B.I. Halperin, C.M. Varma, Phil.Mag. 25,1 (1972)
/6/   S. Hunklinger, A.K. Raychauduri, in "Progress in Low Temperature Physics", Vol. IX, p. 265, D.F. Brewer Ed., North Holland (1985)
/7/   J.R. Klauder, P.W. Anderson, Phys.Rev. 125,912 (1962)
/8/   T.L. Reinicke, Solid State Commun. 32,1103 (1979)
/9/   J. Friedrich, A. Blumen, Phys.Rev. B32,1434 (1985)

EXCITON TRANSPORT IN NAPHTHALENE-DOPED POLY (METHYLMETHYL-ACRYLATE): SPATIAL AND ENERGETIC DISORDER

E. Irene Newhouse and Raoul Kopelman

Chemistry Department
University of Michigan
Ann Arbor, Michigan   48109

ABSTRACT

A novel method is presented for the study of spatially and/or energetically disordered domains in a photoactivated polymeric material. Long-lived phosphorescence and delayed fluorescence spectra and their decay dynamics are studied in a "plexiglass" sample that is both mechanically improved and optically activated via doping with aromatic molecules (naphthalene, anthracene). The dopants are aggregated in sub-microscopic domains ("cavities", "pores"). Fractal-like energy transport and exciton recombination (fusion, annihilation) are observed and related to the structural aspects and the energetic disorder. Energy trapping is studied for samples doped heavily with naphthalene and lightly with anthracene. Exciton fusion is studied on samples with naphthalene dopant (and photo-products: excimer, radial, etc.). While the spectra reveal only the high degree of energetic disorder, the decay dynamics characterize the geometric disorder. The naphthalene singlet exciton transport is about three orders of magnitude slower than in crystalline naphthalene and obeys Stern-Volmer kinetics. This corresponds to normal (non-fractal, non-dispersive) diffusion of the excitation (due, in part, to longer-range transition-dipole-transition-dipole interactions). However, the triplet exciton motion and reaction (fusion) clearly show anomalous (dispersive) diffusion and non-classical (fractal-like) recombination kinetics. The effective spectral dimension is about 1.5 over a wide temperature range (77-165K), indicating that the geometrical disorder ("fractal nature of the cavities") appears to dominate (for the short-range, exchange-like interactions). The triplet exciton transport and kinetics are at least two orders of magnitude slower than in perfectly crystalline naphthalene domains (and all exciton motion is frozen at about 4.2K). Large inhomogeneous fluorescence line-broadening is observed ($\sim$100 cm^{-1}), as well as spectral diffusion (due to triplet exciton energy relaxation).

INTRODUCTION

Interest in the study of energy transport in amorphous media has been stimulated recently by the application of a fractal model to the problem. (See papers by Kopelman et. al and by Hanson et al. in this volume). It is necessary to test such models on experimental systems. Polymer backbones have been assigned fractal properties.[1] One might expect the spaces between them ("cavities", "pores") to also be fractal-like, and thus dopants in a polymer may be forced into a fractal environment. We have chosen naphthalene/poly-(methylmethacrylate) (PMMA) ("plexiglass") for such experiments, as the spectroscopic properties of naphthalene in other phases are very well known, and thus meaningful comparisons can be made between the new data to be presented here and those previously reported for other systems of naphthalene.

EXPERIMENTAL

Zone-refined and potassium-fused naphthalene (Aldrich) was used. PMMA (Polysciences, "medium molecular weight") was reprecipitated from dichloromethane (Baker, spectrograde) by hexane (Baker, hplc grade), and vacuum dried. Samples prepared by solvent casting from this polymer behaved the same as samples made from unreprecipitated polymer. Other samples were made by polymerizing freshly distilled methylmethacrylate (Scientific Polymer Products) with an appropriate amount of naphthalene dissolved in it. 2,2-azodi-isobutyronitrle (AIBN) was used as initiator and the evacuated sealed tube heated to 60C for 24-36 hours. Spectra of these samples were indistinguishable from those made by solvent casting. PMMA is a brittle polymer. These samples, however, are not, due to the plasticizing effect of the dopant.

All spectroscopic measurements were made in an immersion cryostat at 77K unless otherwise noted. This was done because the samples turned brown after about 1/2 hour ultraviolet illumination at room temperature. Excitation was with a 1600W xenon arc lamp, appropriately filtered.[2] A custom-built (Karl Lambrecht Corp.) relay lens was used to match the f number of the 1m Jarrell-Ash monochromator. A cooled ITT F-4013 phototube and PAR 1120 photon counter were used to detect the emission at right angle to the excitation. Spectra were recorded on an LSI 11/03 microcomputer. Phosphorescence and delayed fluorescence decays were recorded with the same optical train using out-of phase shutters in the excitation beam and in front of the monochromator. Only the first half of the monochromator was used, as high rejection of scattered light was not necessary. Detection was with an EMI 9781R phototube and PAR 4202 signal averager, whose output was stored on the microcomputer. Fluorescence decays were excited with the frequency doubled output of a Thodamine 6G solution pumped with a Molectron UV1000 nitrogen laser. Decays were detected with the EMI phototube and a PAR gated integrator (model 164) and boxcar averager (model 162). Due to electrical noise from the laser, the microcomputer could not be used. Instead, decays were recorded with a chart recorder and subsequently digitized.

RESULTS AND DISCUSSION

The fluorescence spectra of naphthalene/PMMA films at 77K are shown in Figure 1. The peak positions are blue-shifted compared to naphthalene in durene[3] at 20K, but this shift is of the order of the full width at half maximum (FWHM) of 19A ($1980cm^{-1}$). At 4K the lines narrow to 12A ($120\ cm^{-1}$), and shift (to the blue) by $20\ cm^{-1}$. It should be noted

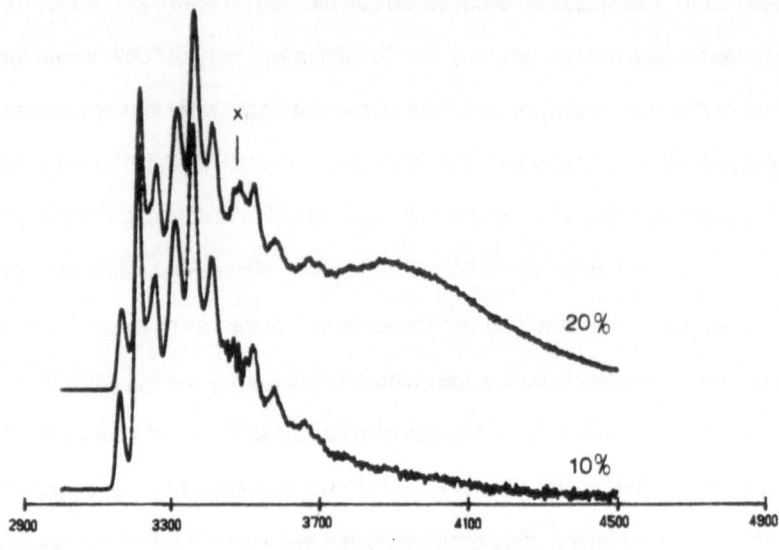

WAVELENGTH(ANGSTROMS)

Fig. 1.   Fluorescence spectra of naphthalene-doped PMMA taken at 77K. The
weight percent of dopant is given next to the lines. (The region labelled
with x marks the region of residual excitation from the xenon arc due to
imperfect overlap of the absorption edges of the filters used. A spectrum
of the light scattered by a pure PMMA sample has been used to correct for
this, but the spectrum may still be slightly distorted in this region).

that the naphthalene crystal Davydov splitting at 4K is of the order of the PMMA sample emission line width[4], and the peaks occur in the same region as those of the amorphous material in these films; thus, microcrystals cannot be detected by their emission. These films are optically clear, and there is no evidence for birefringence in a polarizing microscope. The singlet lifetime is 125 ± 10 nsec, independent of naphthalene concentration. This agrees with the value obtained for crystals.[5]

In the 20% naphthalene sample there is excimer emission peaking at 3900A, which is somewhat higher in energy than the excimer in other systems.[6] The excimer lifetime[7] is 230 ± 15 nsec.

A set of samples lightly co-doped with anthracene was used to determine if singlet migration was occurring in the naphthalene aggregates. The naphthalene singlet lifetime seems to obey Stern-Volmer kinetics in these samples. That is[8]:

$$T(N_A)^{-1} = (\tau^0)^{-1}(1+\kappa N_A)$$

where N_A is the acceptor concentration in moles acceptor per mole donor, $\tau(N_A)$ is the donor lifetime in the presence of a given acceptor concentration, τ^0 is the donor lifetime in the absence of acceptor, and κ is the number of times an exciton hops in the natural lifetime of the donor. κ/τ^0 is thus an effective rate constant for the trapping process. Stern-Volmer plots are given for 20% naphthalene and for 3.7% naphthalene in Figure 2. The extremes in the measured lifetimes are 50 to 120 nsec for the 20% samples, and 70 to 110 nsec for the 3.7% samples. The effective rate constant is 7×10^8 sec^{-1} for the 3.7% sample, and 2×10^9 sec^{-1} for the 20% sample. (The axes in Fig. 2 are according to the formulation of Powell and Soos[8]). We find that anthracene is about 500 times less efficient a quencher in naphthalene/PMMA than in naphthalene crystals.[9] Using an average spacing of 5.5A between molecules in 20% naphthalene (formula from Chandrasekhar[10]), and assuming that diffusion is the rate-limiting step (and cubic symmetry), one can estimate[6] a diffusion length of about 40A.

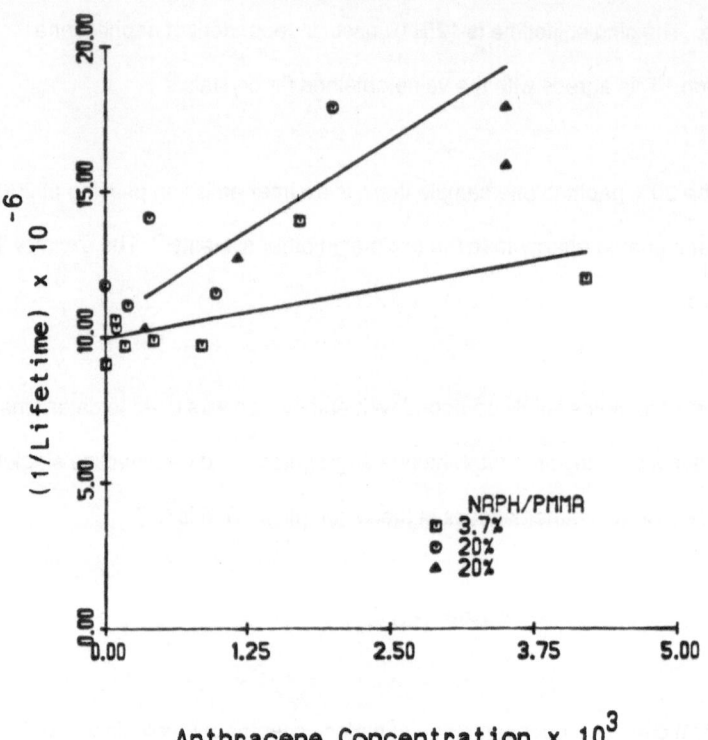

Fig. 2. Stern-Volmer plots of naphthalene singlet lifetimes in anthracene-doped films for 20% and 3.7% naphthalene. The two triangles at the extreme right represent two measurements on the sample and indicate the amount of error encountered by using a fluctuating light source (Molectron laser) with a boxcar integrator as the lower limit of the measured lifetimes is reached.

In contrast to the fluorescence spectrum (singlet emission), the phosphorescence spectrum (triplet emission) is concentration dependent, as is evident from Figure 3. There is a red shift and broadening of the spectrum as the concentration increases. At 4K, the emission spectra of all samples are almost indistinguishable, and resemble the spectrum of the 1% sample at 77K. The FWHM in the 5% sample is about 650 cm^{-1} at 77K.

Peaks near 5250A (19000 cm^{-1}), 5360A (18700 cm^{-1}), and 5800A (17200 cm^{-1},[11] especially prominent in the emission of the 5% sample, are due to the 1-hydronaphthyl radical which is generated by the ultraviolet irradiation of the sample.

The phosphorescence lifetime in samples doped only with naphthalene is concentration dependent (see the next two figures and Table I). If one selects the detection wavelength of the phosphorescence in the 20% sample, the observed lifetime increases the more toward the blue one observes, as evident in Table II. Thus, the emission spectrum represents a distribution of sites with varying intermolecular interactions or a distribution of domain sizes. Some sites, with limited excitation exchange interaction with other sites, have an emission spectrum characteristic of dilute naphthalene systems, and a long emission lifetime. In other regions of the sample there is aggregation of varying degrees, resulting in a red-shifted emission spectrum, and a shorter lifetime, because of triplet-triplet annihilation and/or trapping. The phosphorescence lifetime of all these sample lengthens at 4K (see Table I), implying that triplet energy transport is a thermally activated process that has ceased at 4K.

The excimer emission tail, seen in the most concentrated samples, does not perturb these measurements: the amount of excimer emission present at the time scales used (see Figures 4 and 5) at 4000A (peak of the excimer emission) is less than one-tenth the over-all emission at 5000A (excimer plus triplet emission). Thus the observed excimer emission is largely prompt emission that has died away by the time the detection of the

Table I: Naphthalene triplet lifetimes (5000A)

Concentration (%)	τ (4K) (sec)	τ (77K) (sec)
20	1.9	0.8
10	2.2	1.0
5	2.1	1.7
1	2.3	2.0

Table II: Triplet lifetime as a function of wavelength (77K)

Wavelength (A)	τ (sec)
4800	0.8
4750	0.9
4700	1.6

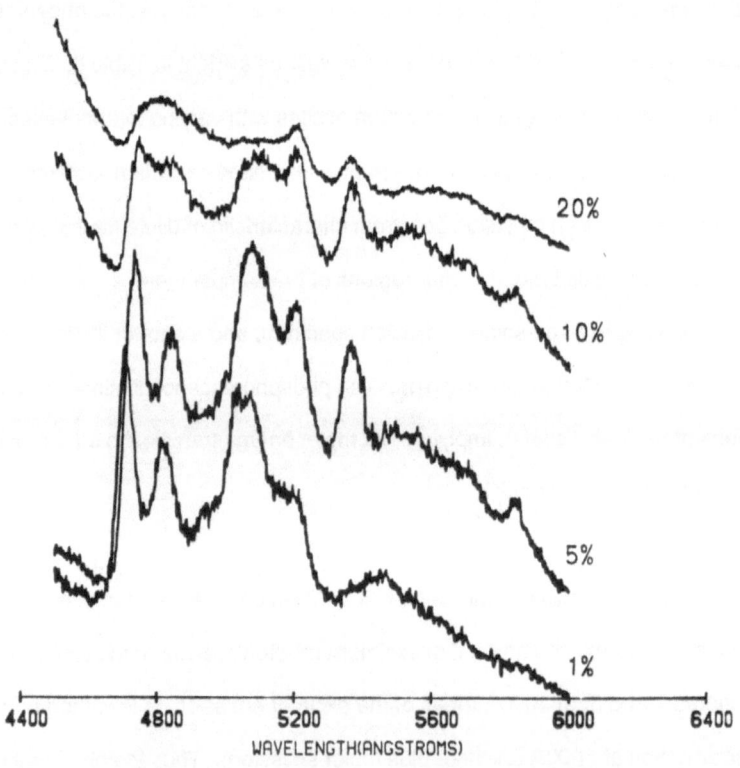

Fig. 3.   Phosphorescence of naphthalene/PMMA films at 77K as a function of naphthalene concentration.

the delayed emissions commences, and thus the delayed emission at 5000A can be effectively assigned to the phosphorescence.

The phosphorescence decay is "bi-exponential", (Figures 4 and 5) with the lifetimes of the two "components" indicated on the Figure. Note also that the deviation from a single exponential (see Figures 4 and 5) decreases with concentration. The phosphorescence decay from a 5% sample can be fit with a single exponential.

While the delayed fluorescence can also be fit to two exponentials, one usually expects a linear $I^{-1/2}$ vs. time behavior at short times and an exponential decay at long times with a decay time 1/2 that of the phosphorescence decay time.[12] Note that the slow component of the delayed fluorescence decays much more rapidly than at twice the slow phosphorescence decay rate. This can be explained as being due to the presence of domains with varying amounts of naphthalene-naphthalene interactions, with triplet-triplet annihilation occurring only on those domains with the strongest interactions (shortest lifetimes). (So far, we are describing the system as an ensemble of different regions, but all with Euclidean topology).

On the other hand, one expects annihilation on a fractal to have a lifetime behavior reflecting the spectral dimension,[13] d_s:

$$I_{df} \propto \rho^x,$$

where I_{df} is the delayed fluorescence intensity, ρ is the exciton density, and $x=1+(2/d_s)$. Specifically the ratios τ_p/τ_{df} observed in these samples correspond to $d_s \sim 1.4$ (18% naphthalene) and $d_s \sim 1.2$ (8.4% naphthalene).

We were interested to check for further evidence of sample heterogeneity.[14] A plot of $\log(I_{df}/I_{ph}^2)$ vs. $\log$(time) (Fig. 6) shows that these samples are heterogeneous, as

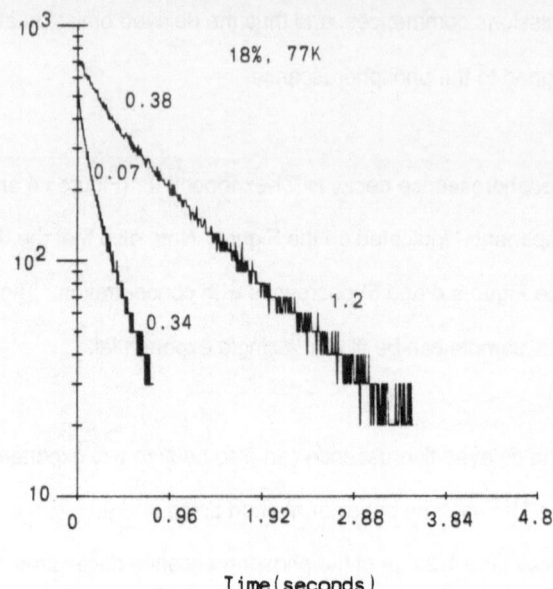

Fig. 4. Phosphorescence and delayed fluorescence decays for 18% naphthalene samples at 77K. The upper curve is phosphorescence, the lower fluorescence. The numbers are the decay time values obtained from a least squares fit to a bi-exponential function. The "steps" reflect the voltage resolution of the A/D converter used to record the data.

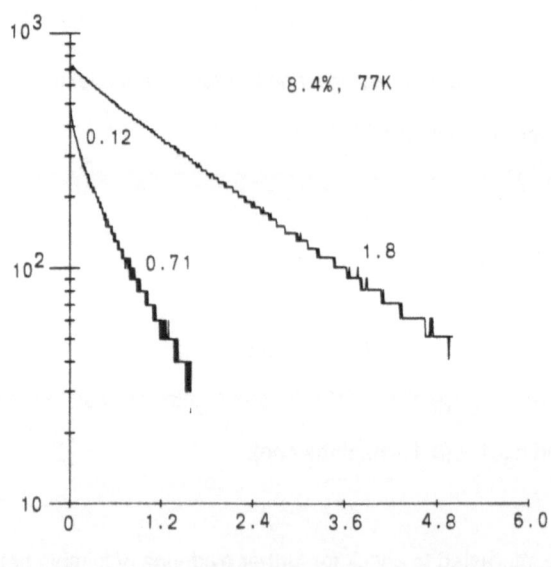

Fig. 5. Phosphorescence and delayed fluorescence decays for 8.4% naphthalene samples at 77K. This figure is completely analogous to Fig. 4.

expected from the phosphorescence decays. (For homogeneous samples these plots are horizontal lines). The "slopes" are 0.2-0.3, corresponding to d_s of 1.6-1.4, for all the data shown in this Figure. A plot of the type shown in Fig. 6 will be linear only as long as the excitons can meet and annihilate. Thus, for finite clusters, one expects to see deviations from linearity as in Fig. 6, when the number of clusters with more than one exciton on them begins to decline significantly. This effect has been demonstrated by simulation.[15]

These data were obtained at high excitation intensity levels. At very low excitation intensity levels, the slope tends toward the horizontal, perhaps because only those excitons with excess thermal energy, to which the medium appears more homogeneous, annihilate at all. Another possible explanation is that, at low exciton densities, annihilation occurs only in those regions of the sample that transport the best, i.e. classically. 5% samples behave similarly to the more concentrated samples at all excitation intensity levels.

If one plots $I^{-1/2}$ vs. time for delayed fluorescence, one obtains a linear plot as shown in inset a in Figure 7. (This is true when triplet-triplet annihilation does not perturb the exciton population (low intensity limit), and in homogeneous media. It is not easy to decide whether one is in the typical low intensity regime if not all sites permit annihilation; thus the mere observation of straight lines in all the inverse root intensity plots was taken as evidence for the achievement of such a low intensity regime). The intercept of such a plot is $I_o^{-1/2}$, where I_o is the initial delayed fluorescence intensity. The value of the slope of this plot depends on the excitation intensity level. If one obtains data at several intensities, a plot of these slopes vs. $I_o^{-1/2}$ yields a line whose slope is the phosphorescence decay rate, i.e. $1/\tau$, where τ is the phosphorescence lifetime, and whose intercept is $\gamma^{1/2}$, where γ is the annihilation rate constant.[12] (The slopes of the least squares lines fitted to these data are within a factor of three of the values measured from the phosphorescence decays, indicating reasonably accurate conversion of the measured decays of an absolute intensity scale). The values of the annihilation rate constant γ are 2×10^{-14} cm^3 sec^{-1} for 18% and 8.4% naphthalene, and 3×10^{-16} cm^3 sec^{-1} for 5.2%

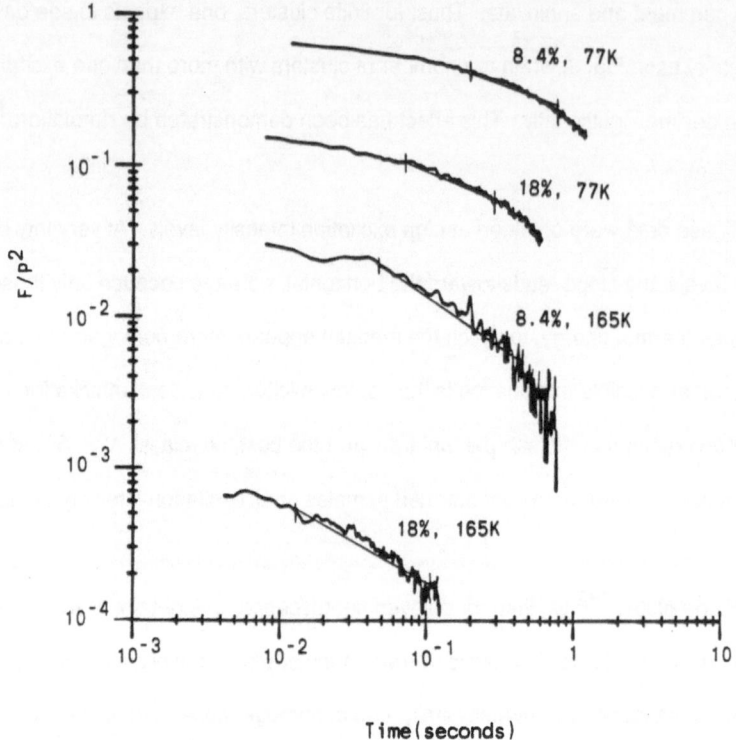

Time(seconds)

Fig. 6.   Test for sasmple heterogeneity effects on triplet transport at two

temperaturess and two concentrations. The concentrations and

temperatures are given on the Figure. The superimposed lines indicate the

slopes of the fit and are plotted only over the region used for the fit. The

values range from 0.2 to 0.3. The kinetics are clearly nonclassical, as a

horizontal line is expected for classical kinetics, and is clearly not observed

here. The values of the ordinate are arbitrary, but the time in seconds can

be read directly from the abscissa.

naphthalene in PMMA. The errors propagate in such a way that these values are upper limits. The more concentrated samples have apparently aggregated to a common structure. The annihilation rate constant for naphthalene crystals, obtained from magnetic field effect measurements[16], is 1.3×10^{-12} cm^3 sec^{-1}. Thus, the aggregate structure seen in the glassy samples differs from the naphthalene crystal structure, which was already evident from the millisecond time scale of the delayed fluorescence in these samples (the time scale for crystals is microseconds or less[17]), and the heterogeneity of the kinetics.

Although the addition of anthracene does result in a slightly decreased naphthalene phosphorescence intensity, the long component of the naphthalene phosphorescence lifetime is unchanged. This is probably because the long component reflects the kinetics of naphthalene molecules isolated from interactions with other molecules. The anthracene delayed fluorescence, on the other hand, is anthracene-concentration dependent, as shown in Fig. 8. (While the lifetime of anthracene in PMMA is ~10 nsec, our decays occur over many milliseconds and must thus be fed from the naphthalene triplet population). This concentration dependence may reflect lifetime quenching by anthracene in those regions of the sample that do transport excitons, and should be reflected in a modest decrease in the lifetime of the short component of the naphthalene phosphorescence decay. This shortening was not detected, probably because of the lack of sensitivity of the fitting technique to modest changes in the fast component of the napthalene phosphorescence decays. In a 3.5% naphthalene sample, the anthracene emission is prompt fluorescence, indicating that triplet transport has ceased (at 77K).

$Log(I_{df}/I_{ph}^n)$ pots for anthracene delayed fluorescence and naphthalene phosphorescence are given in Figure 9 for n=1 and Figure 10 for n=2. At the lowest anthracene levels, it is not clear whether second-order or pseudo-first-order kinetics apply, but at the highest anthracene level reported, it seems that pseudo-first-order kinetics do occur. This level of anthracene is near that at which the anthracene emission spectrum

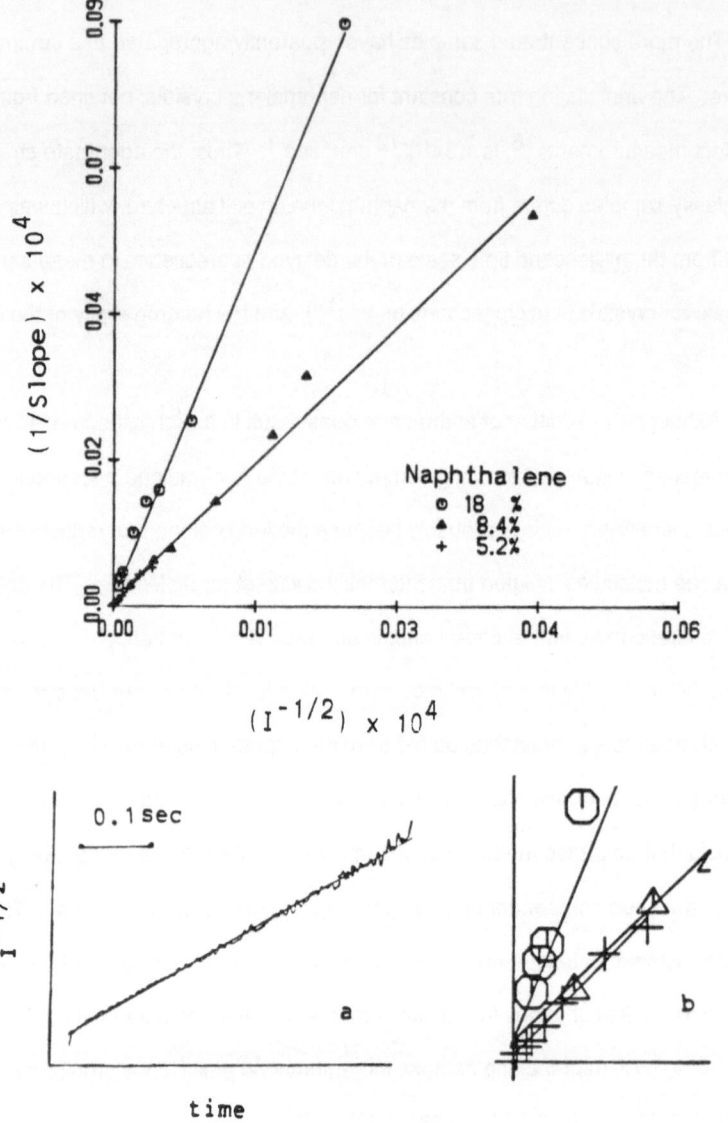

Fig. 7. Determination of the triplet-triplet annihilation rate constant γ for various naphthalene concentrations at 77K. Inset a shows a plot of a delayed fluorescence decay plotted as $I^{-1/2}$ vs. time. Each point on the major part of the figure has as its ordinate the inverse square root of the slope of a plot such as that of inset a, and as its abscissa, the y-intercept of such a plot, i.e., the inverse square root of the initial intensity. both these quantities depend on the excitation intensity, so that each line consists of the results of a series of decays obtained at different intensities. Inset b is a blow-up of the region about the origin of the major part of this figure.

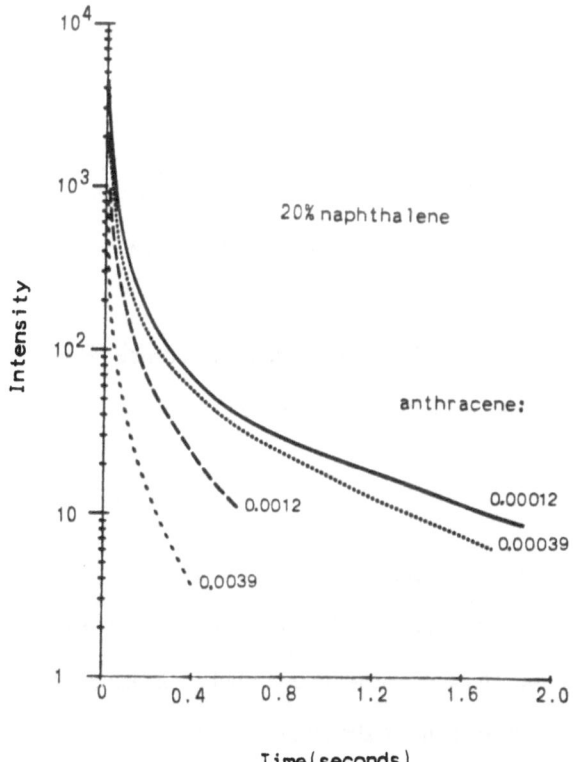

Fig. 8. Anthracene delayed fluorescence decays in naphthalene co-doped PMMA films at 77K. The napthalene concentration is 20%, and the anthracene concentration, in moles anthracene/moles naphthalene, is given on the Figure next to the appropriate curve. The decays have been normalized to the same initial value for comparison purposes.

changes to that characteristic of anthracene dimers, so this trend cannot be studied at higher anthracene concentrations.

SUMMARY AND CONCLUSIONS

The fluorescence emission from these samples is broad, indicating a wide distribution in site energies in this polymeric medium. The lifetime is the same as for naphthalene crystals.

In concentrated films, excimer emission, which is blue-shifted compared to other media, is seen. The excimer emission is greatly reduced at 4K.

The effect of anthracene doping on prompt fluorescence follows a Stern-Volmer rate law and indicates that singlet energy transport is ~500 times less efficient than in naphthalene crystals.

The phosphorescence spectrum of each concentrated sample represents a superposition of spectra from sites with different environments, or with different domain sizes. In the phosphorescence of concentrated samples, "spectral diffusion" of excitons to sites with low energy environments is seen (triplet exciton energy relaxation). The fluorescence and phosphorescence spectra both reflect the energetic disorder in the naphthalene.

The phosphorescence lifetime is also concentration dependent, unlike the singlet lifetime. Triplet transport is a thermally activated process which ceases at 4K.

The non-exponential nature of the phosphorescence decay may indicate that there is a distribution of sites. This is further evident from the abnormally fast decay of delayed fluorescence, which implies that only those sites with shorter phosphorescence lifetimes

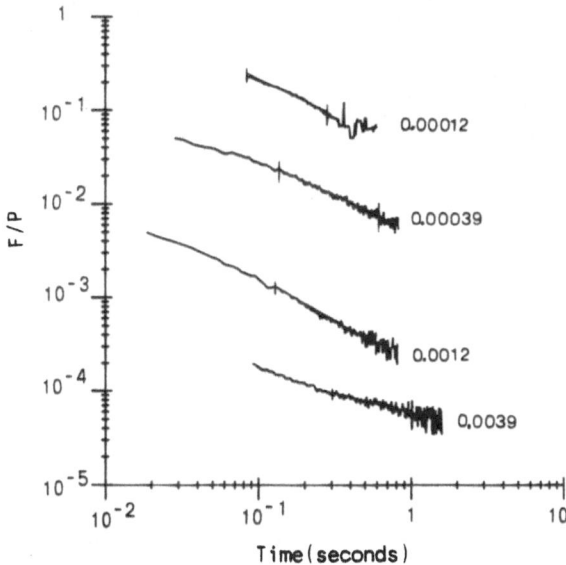

Fig. 9.   Log(anthracene delayed fluorescence + naphthalene phosphorescence) vs.

log(time) for a 20% naphthalene sample at 77K. In the ordinate lable, F

stands for anthracene delayed fluorescence intensity, and P for

naphthalene phosphorescence intensity. The amount of anthracene is

indicated next to each curve.

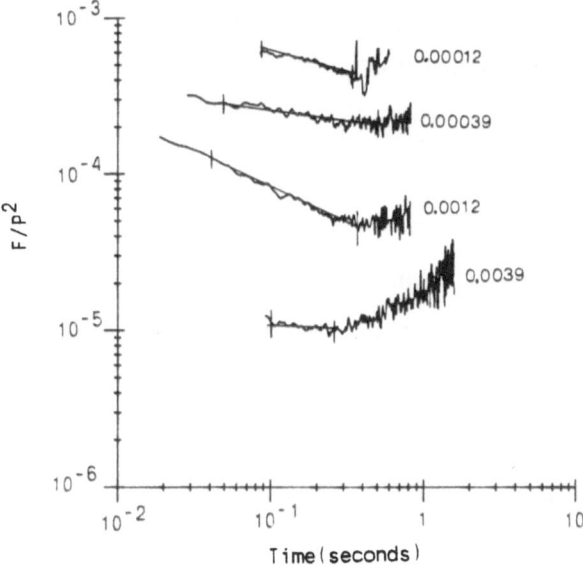

Fig. 10.   Log(anthracene delayed fluorescence + (naphthalene phosphorescence)^2)

vs. log(time) for a 20% napthalene sample at 77K. This Figure is a

companion to the previous Figure.

allow triplet-triplet annihilation. These observations may also be explained by describing the system as fractal-like with a spectral dimension of about 1.5. It has been noted that the behavior of samples consisting of regions with different Euclidean geometry may be fractal-like[18] (provided that they contain one-dimensional bottlenecks).

The delayed fluorescence data can be used to extract an annihilation rate constant which is at least ~100 times smaller than in naphthalene crystals, indicating that the regions supporting triplet migration are not crystalline, in spite of the fact that they seem to have a well-aggregated structure in the two most concentrated samples studied.

$Log(I_{df}/I_{ph}^2)$ vs. log(time) plots demonstrate the heterogeneity by deviating drastically from the classical picture. Forcing a fit with a fractal model yields slopes of 0.2-0.3 and d_s of about 1.3-1.6. The kinetics are thus seen to reflect the geometric disorder, yielding information complementary to the emission spectra, which reflect primarily the energetic disorder.

Singlet exciton transport, probably due to the long-range nature of transition-dipole-transition-dipole interactions, occurs in a manner characteristic of an effectively homogeneous medium in naphthalene-doped PMMA. However, the transport is much slower than in naphthalene crystals.

For triplet energy transport, however, the medium is effectively heterogeneous, with some parts apparently not transporting energy at all. Those regions which do transport do so much less readily than pure crystals. The transport topology is fractal-like, but only as a first approximation.

ACKNOWLEDGEMENT

This research was funded by NSF Grant No. DMR 8303919.

REFERENCES

1. B.B. Mandelbrot, "The Fractal Geometry of Nature", Freeman, New York, **1982,** pp. 227, 329.

2. M. Kasha, J. Opt. Soc. Am., **1948,** 38:929.

3. D.S. McClure, J. Chem. Phys., **1954,** 22:1688.

4. D.P. Craig and S. H. walmsley, "Excitons in Molecular Crystals", Benjamin, New York, 1968, D.P. Craig, L.E. Lyons, and J.R. Walsh, Mol. Phys., **1961,** 4:97.

5. R.P. Parson and R. Kopelman, Chem. Phys. Lett., **1982,** 87:528.

6. S. Ito, M. Yamamota, and Y. Nishijima, Bull. Chem. Soc. Jpn., **1981,** 54:35.

7. The excimer lifetime of naphthalene-containing systems is often longer than the monomer lifetime. See Table 7.3 in J. Birks, "Photophysics of Aromatic Molecules", Wiley, New York, **1970,** p. 351, as well as Table 8 of S.N. Semerak and C.W. Frank, Adv. Polymer. Sci. **1984,** 54:31.

8.   R.C. Powell and Z. Soos, J. Luminescence, **1975,** 11:1.

9. H. Auweter, Ph.D. Thesis, University of Stuttgart, **1978.**

10. S. Chandrasekar, Rev. Mod. Phys., **1943,** 15:1.

11. K. Nakagawa and N. Itoh, Chem. Phys., **1976,** 16:461, J. Martin-Bouyer and P. Martin-Bouyer, J. Chim. Phys. Physicochim. Biol., **1974,** 74:1679.

12. S. Singh, W.J. Jones, W. Siebrand, B.P. Stoicheff, and W.G. Schneider, J. Chem. Phys., **1965,** 42:330.

13. L.W. Anacker, R.P. Parson, and R. Kopelman, J. Phys. Chem., **1985,** 89:4758.

14. R. Kopelman, P.W. Klymko, J.S. Newhouse, and L.W. Anacker, Phys. Rev. B, **1984,** 29:3747.

15. J.S. Newhouse, unpublished.

16. L. Altwegg and I. Zschokke-Granacher, Phys. Rev. B, **1979,** 20:4326.

17. S.T. Gentry and R. Kopelman, J. Chem. Phys., **1984,** 81:3014.

18. C.L. Yang, P. Evesque, and M.A. El-Sayed, J. Phys. Chem., **1985,** 89:3442.

EXCITATION TRANSPORT KINETICS IN VAPOR-DEPOSITED NAPHTHALENE

Laurel A. Harmon and Raoul Kopelman

Department of Chemistry
The University of Michigan
Ann Arbor, Michigan   48109

ABSTRACT

Transport and reactions in heterogeneous, disordered or low-dimensional media are characterized by non-classical kinetic behavior. Spectroscopic and kinetic studies of structurally and energetically disordered naphthalene are presented. The samples, prepared by vapor deposition at low temperatures, are shown to be heterogeneous and to consist of two types of domains: nearly crystalline (type I) and highly disordered (type II). Naphthalene singlet excimers and 1-hydronaphthyl radical are produced in type II regions. The temperature dependence of the steady-state fluorescence spectrum between 5 and 85 K is used to show that thermally-activated singlet exciton transport occurs in regions of both types, as well as from type I to II. We have analyzed the kinetics of triplet-triplet annihilation in type II domains through delayed fluorescence decays. The kinetics are shown to be non-classical and described by a time-dependent rate coefficient of the form $k(t) \propto t^{-h}$ with $h \simeq 0.5$.

INTRODUCTION

Understanding of the kinetics of excitation transport in disordered and/or heterogeneous condensed phases has become increasingly important in a number of fields, such as heterogeneous catalysis and in vivo biological reactions. Organic materials deposited from vapor at low temperatures form non-equilibrium structures with both structural and energetic disorder[1], and, as such, provide excellent model systems in which to study the effects of disorder on transport. Naphthalene is a particularly useful material for this type of work. It has been extremely well-characterized in the crystalline state and its weak intermolecular interactions result in realtively narrow, disorder-induced bandwidths, allowing for more detailed spectroscopy than anthracene or tetracene. We present here spectroscopic studies of vapor-deposited naphthalene from which we draw conclusions regarding the degree of disorder and heterogeneity in this material. By analyzing the kinetics of triplet-triplet exciton annihilation we are able to demonstrate that disorder induces non-classical kinetics, characterized by a time-dependent reaction rate coefficient.

EXPERIMENTAL METHODS

Naphthalene of extremely high chemical purity was necessary in this work to isolate the effects of deliberately introduced structural and energetic disorder. Commercially obtained naphthalene was zone-refined (100 passes), fused with potassium metal, and subsequently zone-refined (500 passes). Only the center third of material thus prepared was used for experiments. This purification procedure has been described in detail elsewhere.[2]

Disordered naphthalene samples were prepared by evaporation onto quartz substrates from a minature oven in the sample chamber of a Janis Research Co. liquid helium dewar. The oven was mounted in a teflon cage and the substrate assembly attached to a sample rod. Before deposition, the cryostat, oven and substrate were cooled to 20-30 K, following which the sample chamber was pumped out to 74 cm vacuum (measured with a mechanical gauge). While the substrate was held in the optical path 4 inches below the oven, the oven was heated to about $35^{\circ}C$. The substrate was raised into position directly opposite the oven mouth and held there typically for 2 minutes. The range of oven temperatures used was $20-50^{\circ}C$. During the deposition process, the substrate temperature increased by 20-30 degrees because there is no heat sink in the evacuated chamber. Within the range of oven and substrate temperatures used, no effects were observed on the spectra of the resulting samples. After deposition was complete, the quartz substrate with its vapor-deposited naphthalene was lowered into the optical path, vacuum in the sample chamber was broken with helium gas and the system recooled to liquid helium temperatures.

Samples were annealed by warming the sample chamber to 150K over a 30 minute interval and maintaining that temperature for 5-6 hours. Naphthalene is volatile and sublimes even at cryogenic temperatures. Consequently the thermal energy requisite for annealing also drives the competing sublimation process. Visual inspection of samples in situ sometimes revealed nonuniformity in sample thickness after annealing. Spectra were unchanged by cycling between 100 and 5K. Samples held at 120 K did not appear to anneal significantly.

Excitation was by 310 nm radiation from a 1600 W Xenon arc lamp passed thorough a .25m Jobin-Yvon monochromator with 2mm slits. Emission spectra were collected through a Jobin-Yvon lm double monochromator with an EMI 9816QB photomultiplier and digitized with a PAR model 1109 photon counter. To obtain delayed fluorescence decays, the excitation beam was shuttered. The delayed emission was passed through two Corning 7-54 glass filters which transmit the UV and eliminate phosphorescence. Time-resolved intensities were monitored by an EMI 9781R photomultiplier whose output was averaged with a PAR model 4202 Signal Averager. The output photomultiplier was protected from lamp emission and prompt fluorescence by a shutter in the emission beam triggered to open 3 msec after the excitation shutter closed. Each decay consists of 1024 points at 1 msec intervals, typically averaged over 100 scans. Scans were repeated at intervals of 10 or 20 seconds to allow the triplet excitations to equilibrate during and after illumination. All experimental data were recorded on an LSI-11/03 laboratory computer; fitting of experimental delayed fluorescence decays was performed on an Ahmdal 470 at The University of Michigan.

SPECTROSCOPY OF VAPOR-DEPOSITED NAPHTHALENE

Emission spectra of an unannealed pure naphthalene sample, recorded at temperatures between 4.5 and 85 K, are presented in Figure 1. The conditions of excitation and collection were kept unchanged throughout the course of the experiment so that intensities may be compared quantitatively. The regions in which emission from the singlet, singlet excimer and triplet dominate are labelled. The single sharp peak at 18550 cm^{-1} is due to 1-hydronaphthyl radical (1-HNR), identified by comparison with previously reported spectra.[3] As seen here, the phosphorescence from pure samples, even at the lowest temperature of this study (4.5K), is extremely weak and broad. The overall lack of triplet emission is due to depletion by triplet-triplet annihilation.

At 4.5K, ordinary fluorescence dominates the emission. The excimer is evident only as a broad region of increased intensity between 26000 and 22000 cm^{-1}. At higher temperatures, the excimer emission develops a pronounced maximum at 23800 cm^{-1}, the position of which is independent of temperature. An additional band at 26500 cm^{-1} in the 85 K spectrum, a shoulder on the excimer band, is most probably emission from pairs of molecules with a slightly different relative orientation than the excimer at 23800 cm^{-1}.[4] with increasing temperature, the excimer and 1-HNR are populated at the expense of the singlet, so that by 85 K the excimer is the major source of intensity. The radical peak is relatively narrow (6 cm^{-1}) and at this scanning rate the peak intensity if distorted. However, increase in 1-HNR intensity relative to the fluorescence is quite clear, as is the fact that the intensity of the excimer at the highest temperature far exceeds that of the radical. We conclude from the relative intensities that excimer-forming sites are present in much higher concentrations than radical-forming sites. This is expected in view of the fact that the production of 1-HNR involves genuine photochemistry.

Fluorescence spectra of a single freshly-deposited sample, recorded between 5 and 20 K, are shown in Figure 2. "O_1" designates the origin of naphthalene fluorescence. At 5 K, the positions and relative intensities of the major peaks correspond closely to

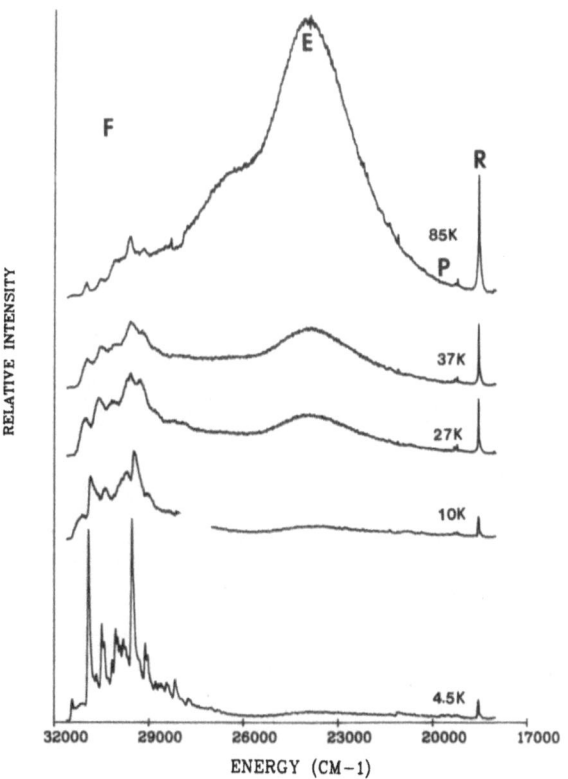

Figure 1.   Emission spectra of freshly-deposited naphtlanele recorded between 4.5 and 87K. (F) Fluorescence, (P) Phosphorescence, (E) Excimer, (R) 1-hydronaphthyl radical

those observed in the fluorescence of pure single crystals at this temperature. The width of the 512 cm^{-1} vibronic band at 30915 cm^{-1} is about 70 cm^{-1} (FWHM) (significantly increased over the single crystal value[5] of 10 cm^{-1}) and is typical of the spectra of these non-crystalline samples. The 5K spectrum, essentially a broadened single-crystal spectrum, will be referred to as Spectrum I.

O_{II} designates a region of increased intensity between the 0-0 and 512 bands (Figure 2) which is not seen in single crystal spectra. This appears in all fluorescence spectra of our vapor-deposited samples, although at low temperatures there is no well-defined maximum. With increasing temperature, the intensity of Spectrum I is reduced

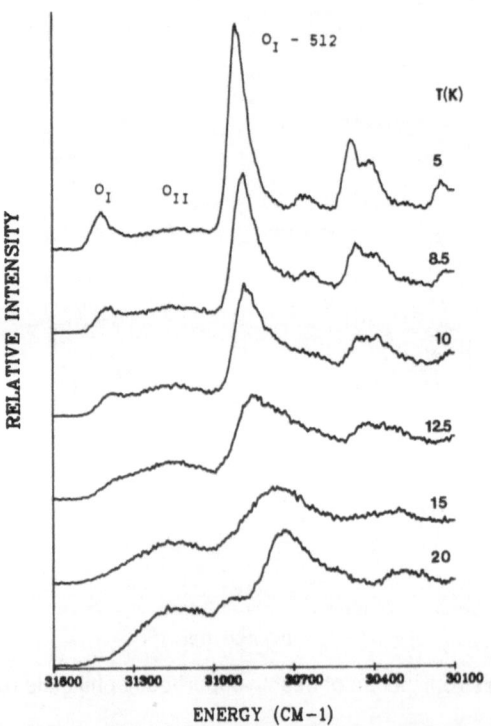

Figure 2.   Fluorescence spectra of freshly-deposited naphthalene recorded between 5 and 20K (See text).

and the maxima shift to lower energy. Simultaneously, the relative intensity of the broad

feature increases and develops a pronounced maximum at 31160 cm^{-1} above about 10K.

There is also a considerable buildup of intensity below each of the bands of Spectrum I,

especially the 512 cm^{-1} band. While the number of overlapping broad bands precludes

the extensive set of assignments possible at 5K, the higher temperature spectrum may be

explained in terms of two distinct sets of naphthalene vibronic bands. One set, Spectrum

I, is essentially like that of the crystal at low temperatures; this set broadens, shifts to

lower energy and is reduced in intensity with increasing temperature. The second set,

which will now be designated Spectrum II, consists of extremely broad bands built on the

31160 cm^{-1} origin. The relative intensity of Spectrum II increases with respect to

Spectrum I as the temperature is increased and its peak positions move to lower energy

above about 10K.

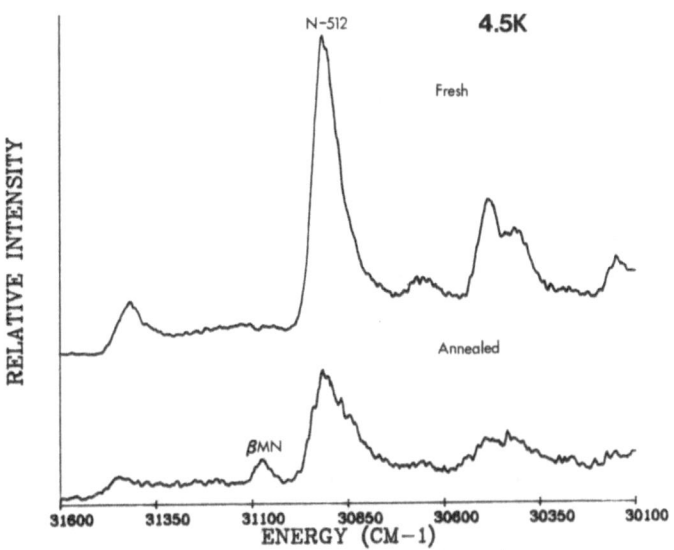

Figure 3.   Fluorescence spectra of vapor-deposited naphthalene before and after
               annealing. βMN denotes the beta-methyl naphthalene origin. Relative
               intensity of the annealed spectrum has been doubled.

In Figure 3, the fluorescence spectrum at 4.5K of a single sample is shown before and after annealing. The relative intensity of the annealed film spectrum has been doubled in this representation. We can see that there has been a considerable reduction of fluorescence intensity after annealing. When the full emission spectrum of the annealed film at 4.5 (Figure 4) is compared to the bottom spectrum of Figure 1, we see a relative enhancement of excimer and 1-HNR emission by annealing. Part of the reduction of fluorescence intensity may therefore be explained by increased trapping at excimer and 1-HNR sites; in addition, the competition between sublimation and annealing processes at elevated temperatures results in sample loss and lower signal levels.

The fluorescence spectrum of an annealed spectrum, recorded at temperatures between 4.5 and 32K, is shown in Figure 5. Fluorescence spectra of the same sample before annealing were presented in Figure 2. The naphthalene 0-0 at 4.5K shows a blue

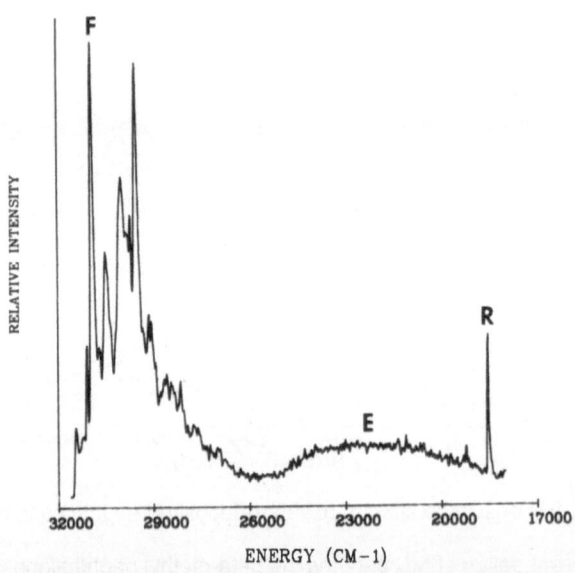

Figure 4.   Emission spectrum after annealing of vapor-deposited naphthalene at 4.5K. See Figure 1 for legend.

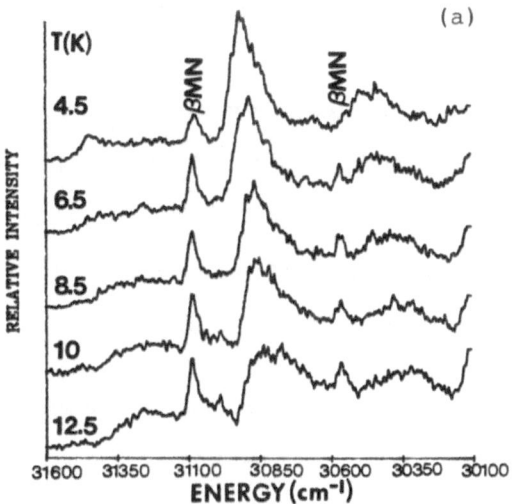

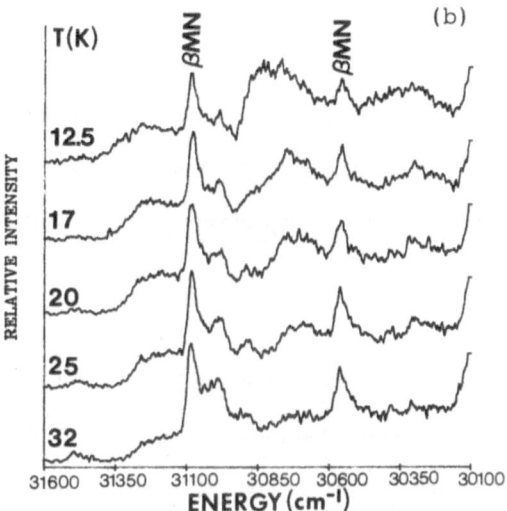

Figure 5.   Fluorescence spectra of vapor-deposited naphthalene after annealing, recorded between 4.5 and 32K. The peaks due to βMN are indicated. a) 4.5 - 12.5K, b) 12.5 - 32K.

shift of about 10 cm^{-1} with respect to the fresh sample spectrum but still lies about 30 cm^{-1} below the single crystal origin. The intensity of the 0-0 peak drops between 4.5 and 8K, then appears to increase steadily with temperature until a peak is again apparent at about 25K. At this temperature the 0-0 band is centered at 31475 ± 5 cm^{-1} which matches the single crystal value[6] of 31476 cm^{-1} at 4K. After annealing, the maximum of the broad band, O_{II}, is also blue-shifted with respect to its fresh counterpart: 31250 cm^{-1} vs. 31160 cm^{-1} in the fresh sample. Its intensity increases up to about 17K and then decreases, together with the remaining naphthalene bands. At 32K, naphthalene emission from the annealed film is very weak and all bands, both those of Spectrum I and Spectrum II, are shifted to higher energies compared to fresh samples.

A peak near 31080 cm^{-1} appears in the spectrum of annealed samples which is entirely absent in the fresh samples. This peak is attributed to a chemical impurity, β-methyl naphthalene (βMN), populated by trapping. At 4.5K the βMN peak occurs at 31075 cm^{-1} and is approximately 60 cm^{-1} wide (FWHM). Between 4 and 8K, βMN intensity increases; above this temperature it remains constant. Between 6 and 32K, the position of the βMN maximum is at 31085 cm^{-1} and remains unchanged with a width of about 30 cm^{-1}. The beak is blue-shifted from the position of βMN in naphthalene-h$_8$ single crystals:[7] 31060 cm^{-1}. The bandwidth is considerably increased over previously reported values: <6 cm^{-1} and 0.3 cm^{-1} in naphthalene-h$_8$ single crystals[6,7]; 8 cm^{-1} in a 64% naphthalene-h$_8$/naphthalene-d$_8$ mixed crystal.[7]

The spectra presented here demonstrate that naphthalene samples prepared by vapor-deposition at low temperature display characteristic features of disordered organic materials:[1] both absorption and fluorescence are red-shifted in comparison to single crystal spectra; linewidths are increased in absorption[2] and emission; the formation of physical traps is observed, prevented by structural constraints in single crystals.

We have concluded that Spectrum I arises from a communicating set of moderately perturbed naphthalene molecules. Site-to-site fluctuations in the energy of the singlet state create local minima which act as shallow traps at low temperature, regardless of their position in the overall energy distribution. The probability of thermalization out of these local traps increases with temperature, thereby increasing the probability of reaching lower energy sites in the distribution. Energy relaxation occurs and the set of occupied site energies no longer reflects the total distribution of site energies: the observed maximum in emission moves to lower energy. With energy relaxation the bands lose their symmetrical shapes, and at equilibrium are expected to be governed by a convolution of the site energy distribution with a Boltzmann distribution. Because the absorption spectrum does not change with temperature,[2] the actual site energies are unaffected, at least in this regime. The red shift of Spectrum I with increasing temperature is therefore attributed to energy relaxation.

Those molecules that give rise to Spectrum II comprise a set of traps with an approximately Gaussian distribution of energies centered about 350 cm^{-1} below the origin of Spectrum I. Direct excitation of these type II sites is ruled out by two observations: 1) both relative <u>and</u> absolute intensities of spectrum II increase with temperature; and ii) the relative intensities of I and II at a single temperature are independent of excitation wavelength.[2] Therefore, type II sites are a minority and populated by energy transfer rather than direct excitation. The relative intensity of Spectrum II compared to Spectrum I is then expected to increase with increasing temperature due to the rise in trapping probability, as is observed.

Energy relaxation evidenced by a red shift is observed in Spectrum II but at higher temperatures than Spectrum I. If we estimate an average trap depth, ΔE, by the difference between the origins of I and II, i.e. 350 cm^{-1}, then at 85K $\Delta E/kt \cong 6.3$; detrapping into I is not a probable mechanism for population redistribution within type II sites. The observed energy relaxation in II must be a product of thermally-activated transport among the type II sites themselves, implying that they are in communication with each other and spatially correlated. The bandwidth therefore reflects disorder within type II domains rather than energy differences between domains. Because these sites are responsible in part for quenching Spectrum I, they must also be in communication with type I sites, and we conclude that they are located between or near the boundaries of regions of local order (type I).

The spectra of Figure 1 are evidence for further thermally-activated transport, on a much wider temperature range, as the naphthalene singlets are trapped by excimer-forming sites and 1-HNR sites. At 85K, $kT \cong 60cm^{-1}$ and, if σ is given by the halfwidth of the napthalene 0-0 as estimated above ($35\ cm^{-1}$), then $kT/\sigma > 1$. Under these conditions the near-crystalline sites (I) are entirely depopulated: the "naphthalene" fluorescence in the 85K spectrum is in fact Spectrum II. Comparison of the 37 and 85K spectra in Figure 1 shows that Spectrum II itself loses intensity to the excimer and radical.

The effect of annealing is primarily to reduce local fluctuations in site energy, thereby improving transport as evidenced by increased trapping at type II, excimer, 1-HNR and βMN sites. In addition, the population of the low concentration βMN impurity after annealing must be a consequence of improved transport. In this respect, annealing is analogous to increased temperature. The thermal energy available at the annealing temperature is sufficient to allow some molecular rearrangement and relaxation toward the equilibrium structure, as seen in the high temperature 0-0 of type I fluorescence (Figure 5(b)) which occurs at the position of the single crystal origin. The distribution of deeper trap energies, type II sites, does not seem to be greatly altered except for the slight blue shift.

The appearance of βMN emission after annealing was entirely unexpected and proved to be persistent from sample to sample. The βMN is not detectable in either absorption of fluorescence spectra of single crystals of this material; its concentration has been placed well below 10^{-6} mole fraction.[7] In contrast, in the spectrum of a sample prepared from naphthalene in which the βMN concentration was estimated to be just 10^{-6} mole fraction, βMN emission dominates the fluorescence after annealing.[8,2]

There are two possible explanations for the increased trapping efficiency of βMN in the highly purified naphthalene samples of this work: segregation (without aggregation) of βMN during sample formation or creation of large energy funnels around the βMN sites.

Figure 6.   Sketch of proposed domain structure of vapor-deposited naphthalene. Unshaded areas represent nearly-crystalline, Type I regions. Shaded areas represent highly disordered, Type II regions. The relative widths of the type II regions have been exaggerated. (R) 1-HNR; (E) Excimer.

Segregation would increase the local concentration of βMN in certain regions (and reduce it in others). This would require considerable mobility of βMN during the condensation process and a driving force for segregation, neither of which seems likely. Little segregation is observed in even casually grown crystals with βMN concentrations below the saturation limit. In the second case, that of funnel formation, a significant number of naphthalene molecules in the vicinity of each βMN molecule are perturbed and lie at lower energy, thereby increasing the effective trap size.

The funnel effect in the singlet state has been studied carefully by Gentry in thoroughly annealed, βMN-doped single crystals of naphthalene, prepared by a modified Bridgeman technique.[9] The influence of βMN was found to extend over a small number of sites, most probably nearest neighbors only. In contrast, the samples here are prepared under non-equilibrium conditions, in which the effect of βMN on its surroundings might be more extensive. This may be particularly true in regions of greater disorder, i.e. type II. It may be that βMN molecules found in type II regions are accompanied by large enough funnels to act as particularly efficient traps once transport has been improved by annealing. Comparison of disordered samples of very dilute naphthalene-h_8 in naphthalene-d_8 to those of βMN in naphthalene-d_8 should provide some insight into the mechanism of this enhanced trapping efficiency because no structural perturbation is expected to be introduced by $C_{10}H_8$ in $C_{10}D_8$.

A tentative description of the physical structure of these samples, consistent with the spectroscopic results, is sketched in Figure 6. Reasonably ordered regions exist in which fluctuations of molecular coordinates from single crystal values are relatively small and random; these regions give rise to spectrum I. Forming the foundaries of such regions are areas of much greater structural disorder, acting as lower-dimensional, energetically-disordered systems, which are characterized by an effective bandwidth of about 100 cm^{-1}, observed in Spectrum II. We observe that the excimer emission does not rise dramatically until thermalization allows transport within regions II and therefore place the majority of excimer-forming sites in type II regions. For similar reasons, the

1-hydronaphthyl radical is also assumed to be produced mainly in type II regions. We note that there is no information regarding the concentration of sites which are capable of excimer or radical formation.

TRIPLET EXCITATION TRANSPORT KINETICS

The kinetics of exciton motion in a disordered system has been analyzed in terms of the number of distinct sites visited by a random walker[10] in an equivalent medium, $S(t)$. For random walks on a fractal structure:

$$S(t) \propto t^f \ , \ 0 \leq f \leq 1 \tag{1}$$

The exponent f has been related to a characteristic dimension of the medium, d_s, the spectral dimension[11], by $f = d_s/2$. The triplet-triplet annihilation reaction rate coefficient, $k(t)$, is related to $S(t)$ by[12]:

$$k(t) \propto \frac{dS(t)}{dt} \ \alpha \ t^{f-1} \tag{2}$$

$$= k_0 t^{-h} \qquad h \equiv f - 1$$

in which $h=0$ or $f=1$ expresses the limit of a time-independent rate coefficient. This form of $k(t)$ has been observed in simulations of relations in energetically-disordered media.[13] Experimental studies of exciton annihilation in substitutionally-disordered naphthalene crystals (near percolation)[14], porous glass and artificial membranes[15] and polymeric glasses[16], have also found a power law dependence of the rate coefficient on time.

We have selectively examined the kinetics of triplet excitation transport in type II regions of our vapor deposited naphthalene by monitoring triplet-triplet exciton annihilation in the millisecond time domain. In crystalline naphthalene, this process occurs on a microsecond or shorter time scale[9]: annihilation observed at longer times is due to transport which has been slowed by the presence of disorder.

The kinetics of triplet-triplet annihilation in structurally disordered naphthalene is examined through analysis of the time-dependence of the resulting delayed fluorescence. We start from a rate equation for the triplet population density, $\rho(t)$, incorporating both annihilation and monomolecular decay:

$$- \frac{d\rho}{dt} = k_{ann} t^{-h} \rho^m + k_1 \rho \quad . \tag{3}$$

In this expression, the molecularity, m, of the annihilation reaction has been left unspecified to include both binary (m=2) an unary (m=1) processes. Although annihilation is a strictly binary reaction, pseudo first-order kinetics may be observed if one excitation belongs to a population whose density is time-independent, as for example a constant population of long-lived excited traps.[9] The form of k(t) from equation (2) has been retained; k_1, the monomolecular decay rate, is equal to the inverse natural lifetime of the excitation in the medium. We note that eqn. (3) contains as a special case an annihilation rate constant if h=0. Equation (3) has been solved for the four possible cases: m=1 or 2, h=0 or h≠0.[1]

The delayed fluorescence intensity, $I_{df}(t)$, is proportional to the first term in (3):

$$I_{df}(t) \propto k_{ann} t^{-h} \rho^m(t) \quad . \tag{4}$$

Explicit expressions for $I_{df}(t)$ are found by substitution of $\rho(t)$ obtained from solutions of (3). The resulting expressions for $I_{df}(t)$ are summarized in Table 1. At early times, these

Table 1. Summary of expressions for $I_{df}(t)$. γ is the incomplete gamma function.

h = 0	
m=1	$k_{ann} \cdot \rho_0 \exp\{-(k_{ann}+k_1)t\}$
m=2	$\left(\dfrac{k_1^2}{k_{ann}}\right)\left\{\left(1 + \dfrac{k_1}{k_{ann}\rho_0}\right)e^{k_1 t} - 1\right\}^{-2}$
h ≠ 0	
m=1	$k_{ann}\rho_0 t^{-h}\exp\left\{-\dfrac{k_{ann}t^{1-h}}{1-h} - k_1 t\right\}$
m=2	$k_{ann}^2\rho_0^2 t^{-h}e^{-2k_1 t}\left\{\dfrac{k_{ann}}{k_1^f}\rho_0\gamma(f,k_1 t) + 1\right\}^{-2}$

Table 2. Comparison of different models in fitting delayed fluorescence.

	k_{ann}	$k_1(\tau)$	h	χ^2
h = 0				
m = 1	6.56	2.38 (0.42)	0	4.82
m = 2	7.61	0.00054 (1851)	0	1.09
h ≠ 0				
m = 1	0.29	1.13 (0.88)	0.49	0.0051
m = 2	0.13	0.65 (1.5)	0.49	0.0051

expressions reduce to t^{-h} (for $h \neq 0$); at long times we find simple exponential behavior with rate constants equal to $k_{ann} + k_1$ (m=1, h=0) or $k_1 m$ (m=1 or 2, h>0).

A simplex procedure[17] has been used to fit experimental delayed fluorescence decays by finding the set of parameters $[(k_{ann}\rho_0), h, k_1]$ in each expression of Table 1 which best describes the experimental decay. The product $(\rho_0 k_{ann})$ was treated as a single parameter. The fitting procedure was found to be insensitive to bias by initial guesses of the parameter values. The exponent, h, proved to be the primary factor in determining the quality of the fit. By fixing k_1 and fitting only $(\rho_0 k_{ann})$ and h, we found that an increase in k_1 from 0.4 to 0.6 sec^{-1} results in (at most) a 15% increase in the calculated value of h. This is reasonable in light of the fact that the decays are fit over only about 0.5τ.

The results of fitting a single decay are summarized in Table 2 and shown graphically in Figure 7. Fits with a time-independent rate coefficient, i.e. h=0, are compared to those in which h is allowed to vary. The magnitudes of the corresponding X^2 values for fits with h=0 suggest that a time-dependent rate coefficient is necessary to describe the experimental decay. This is substantiated by comparison of Figure 7a with Figure 7b. The best fits obtained with h=0, i.e. k(t)=constant, are shown in Figure 7a, with the residuals plotted below. We see here that a time-independent rate coefficient cannot describe the experimental decay. It is clear that a time-dependent rate coefficient ($h \neq 0$) is necessary to reproduce the observed time dependence of the delayed fluorescence. Fits obtained with m=1 and m=2 are indistinguishable in Figure 7b. However, unrealistic values of the natural lifetime are necessary to fit with a unary model (m=1). We therefore conclude that the annihilation process in these regions is primarily occurring between excitations from a single population. We note that the value of h obtained is independent of m, indicating that k(t) is the dominant factor in the decay over this time regime.

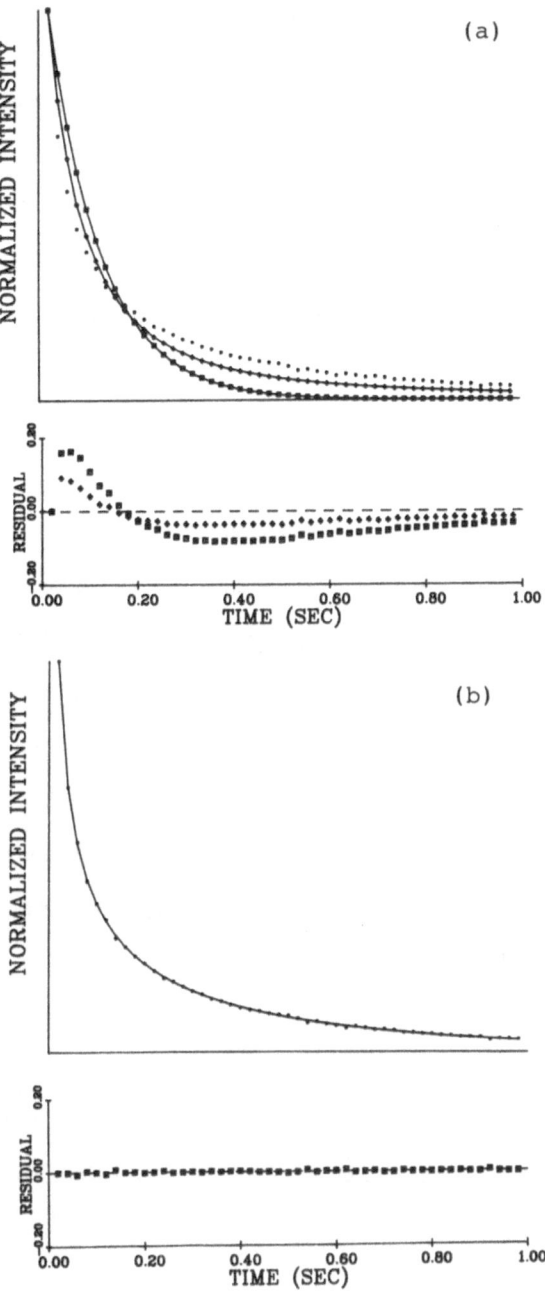

Figure 7.   Delayed fluorescence decay of vapor-deposited naphthalene with fits.

(a) <u>Time-independent rate coefficient</u>. Circles: experimental data. Solid

line with squares: best fit, m=1. Solid line with diamonds: best fit, m=2.

(b) <u>Time-dependent rate coefficient</u>. Circles: experimental data; solid line:

best fit; m=2. Residuals are plotted.

These results were verified by repeated experiments with other vapor-deposited naphthalene samples. All of the qualitative spectral and kinetic features proved to be reproducible, and h was found to vary between 0.45 and 0.6. These value of h are comparable to those obtained from low concentration isotopic mixed naphthalene crystals[14] and are indicative of a highly ramified medium. The values of h were found to be unchanged after annealing; this confirms that the transport observed on this time scale occurs in type II regions since they are affected little by annealing.

CONCLUSIONS

Our work reported here was demonstrated that disordered naphthalene samples prepared by vapor-deposition at low temperatures are heterogeneous and contain both near-crystalline and highly disordered domains. This method has been shown to provide reproducible model systems for the study of the effects of deliberately introduced disorder on excitation transport. We have been able to completely describe the delayed fluorescence, arising from triplet-triplet annihilation in the disordered regions of our vapor-deposited naphthalene samples, by rate expressions incorporating both annihilation and natural decay. The annihilation process in these samples is well-described by a time-dependent rate coefficient, $k(t) \propto t^{-h}$, with h near 0.5.

ACKNOWLEDGEMENTS

This work was supported by NIH Grant No. 2 R01 NS80116-17.

REFERENCES

1.   H. Bassler, Phys. Stat. Sol. B. 107, 9(1981).

2.   L.A. Harmon, Doctoral dissertation, The University of Michigan, 1985.

3. T. Okubo, N. Itoh and T. Suita, <u>Mol. Cryst. Liq. Cryst.</u> 66, 1599 (1977); K. Nakagawa and N. Itoh, <u>Chem. Phys.</u>, 16, 461 (1976); K. Nakagawa and N. Itoh, <u>J. Phys. Soc. Japan</u> 44, 1619 (1978).

4. P.F. Jones and M. Nicol, <u>J. Chem. Phys.</u> 43, 3759 (1965).

5. V.L. Broude, E.F. Sheka and M.T. Shpak, <u>Akademii Nauk SSR Bull. Phys. Ser.</u> 27, 597 (1963).

6. A. Propstl and H.C. Wolf, <u>Z. Naturforsch.</u> 18a, 724 (163).

7. S.T. Gentry, Doctoral Dissertation, The University of Michigan, 1983.

8. P.W. Klymko, Doctoral Dissertation, The University of Michigan, 1984.

9. S.T. Gentry and R. Kopelman, <u>J. Chem. Phys.</u> 81, 3014 (1984).

10. E.W. Montroll and G.H. Weiss, <u>J. Math. Phys.</u> 6, 167 (1965).

11. S. Alexander and T. Orbach, <u>J. de Physique Lett.</u>43, L624 (1982); R. Rammal and G. Toulouse, <u>J. Physique Lett.</u> 44, L13 (1983).

12. P.G. de Gennes, <u>C. R. Acad. Sci. Ser. B</u> 296, 881 (1983); P.W. Klymko and R. Kopelman, <u>J. Phys. Chem.</u> 87, 4565 (1983).

13. L.W. Anacker, R. Kopelman and J.S. Newhouse, <u>J. Stat. Phys.</u> 36, 591 (1984); G. Schonherr, R. Eiermann, H. Bassler and M. Silver, <u>Chem. Phys.</u> 52, 287 (1980); G. Schonherr, H. Bassler and M. Silver, <u>Phil. Mag. B</u> 44, 369 (1981).

14. R. Kopelman, P.W. Klymko, J.S. Newhouse and L.W. Anacker, <u>Phys. Rev. B</u> 29, 3747 (1984).

15. J. Prasad, S.J. Parus and R. Kopelman, this conference.

16. E.I. Newhouse and R. Kopelman, this conference.

17. J.M. Demas, <u>Excited State Lifetime Measurements</u>, Academic Press, New York, New York, 1983.

RESPONSE OF AN INSULATING MATERIAL TO PHOTOEXCITATION

David W. Brown, Bruce J. West[*], and Katja Lindenberg

Department of Chemistry
University of California at San Diego
La Jolla, CA 92093 U.S.A.

ABSTRACT

We focus on the dynamics of polaron formation following photoexcitation of a deformable insulator. A bare exciton is seen to evolve toward a polaron-like state by shedding energy into the vibrations of the medium. The radiated energy propagates away from the excitation region in the form of sound waves, resulting in the emergence of a persistent deformation of the medium about the excitation region.

INTRODUCTION

When an insulating material is excited, *e.g.* by an impinging pulse of laser radiation, it is common for the migration of electronic energy through the material to be observed despite the absence of an electric charge current. In many such materials, and in particular in many molecular aggregates, the phenomenon is well suited to description in terms of the dynamics of Frenkel excitons,[1,2] the elementary excitations of the aggregate corresponding to the excited states of the isolated molecular units. For many years, effort has been directed at understanding the dynamics of particles such as Frenkel excitons in deformable media; that is, media in which the reaction of the medium to the presence of an excitation has significant dynamical consequences. Of particular current interest are such naturally occurring macromolecular systems, as α-helix proteins[3,4] and their quasi-one dimensional solid-state analogs,[5] the myosine molecule involved in muscle contraction,[6] and E. coli DNA in vitro[7,8] and in vivo.[9,10]. These systems are often characterized as "soft-chains" though the free macromolecules may display considerable rigidity. The models most frequently applied to these systems take the isolated medium or chain to be adequately represented in the harmonic approximation, and take the energy of interaction between the particle and the medium to be linear in the displacement field of the medium[11–19]. Such models have spawned a rich literature in which the transport is variously described as stochastic, as characteristic of a random walker;[1,2,14,15] coherent, as characteristic of energy band theory;[1,2,14,15] non-dispersive, as characteristic of solitons;[3,4] etc.. Essential to such descriptions is the concept of the polaron and the "dressing" which distinguishes it from the "bare" particle from which it arises[11–16,19]. Formally, the dressing is effected through a canonical transformation which, when the particle is immobile, diagonalizes the linear-coupling Hamiltonian. In this case the transformation identifies a new particle which is decoupled from the new oscillatory modes of the medium. The new particle, the polaron, is a mixture of the original states in which the bare particle is combined with the deformation which it induces in the lattice. When the bare particle is not immobile, viz. when matrix elements of the Hamiltonian between distinct site-states are non-vanishing, the diagonalization effected by the transformation is

---

[*]Permanent address: Center for Studies of Nonlinear Dynamics, La Jolla Institute, 3252 Holiday Court, Suite 208, La Jolla, CA 92037 U.S.A.

only approximate. Hence, the polaron defined by the canonical transformation provides only an approximate representation of the microscopic state. The non-stationarity of the polaron as a microscopic transport state leads to a complex transport theory which occupies a large part of the relevant literature.[1,2,14,15]

Polaron transport theories tend to reflect one of two qualitatively different points of view: The particle may be considered to be in an extended mobile state and the observed diffusion to be the consequence of scattering out of this extended state. Alternatively, the particle may be considered to be in a localized state and the observed diffusion to be the consequence of events in the lattice which result in transfers between localized states. Some general theories suggest that each perspective may be valid in appropriate parameter regimes, or that both kinds of states may coexist. The question of which qualitative picture is appropriate to a given circumstance, as well as many other questions of fundamental interest, turns on the stability of various transport states and the mechanisms through which unstable states relax.

The premier example of such a relaxation within the polaron problem is the dressing process through which a polaron is thought to form from an initially bare state via vibrational relaxation. It is the viewpoint of the authors that much can be learned about this process even in the absence of transport. By studying the evolution of the many-body state in the immobile exciton limit of the standard exciton transport model, the dressing process can be characterized through exactly calculable quantities. These will allow a clear understanding of the character of the asymptotic many-body state approached by an initially bare exciton, as well as the detail of the dynamics through which the initial state evolves into its asymptotic dressed form.

We develop such exact diagnostics in the following section, where we calculate the bare lattice energy and deformation, and certain inner products of bare and dressed wave functions.

DIAGNOSTICS

The standard model for Frenkel exciton transport employ the linear-coupling Hamiltonian[6]

$$H = \sum_{m,n}(E_m\delta_{mn}+J_{mn})a_m{}^\dagger a_n + \sum_q \hbar\omega_q(b_q{}^\dagger b_q+\tfrac{1}{2}) + \sum_{q,m}\chi_m^q\hbar\omega_q(b_q{}^\dagger+b_{-q})a_m{}^\dagger a_m \quad (1)$$

in which $b_q{}^\dagger(b_q)$ creates (annihilates) a phonon of mode q with frequency ω_q, and $a_m{}^\dagger(a_m)$ creates (annihilates) an exciton in the site-state m having the single particle energy E_m. The coupling coefficient χ_m^q has the property $\chi_m^{q*} = \chi_m^{-q}$, and may be written $\chi_m^q = \chi^q\exp[-iqm]$ in an ordered medium (q and m may be considered vectors in the appropriate number of dimensions, and qm their dot product). The quantity J_{mn} resonantly couples sites m and n.

In the limit $J_{mn} \to 0$, which we consider for the remainder of this section, this Hamiltonian is diagonalized by canonical transformation[11]

$$U = \exp\left[- \sum_{q,m}\chi_m^q(b_q{}^\dagger - b_{-q})a_m{}^\dagger a_m\right] \quad (2)$$

and the operators representing the new, non-interacting excitations are given by

$$A_m = Ua_mU^\dagger = a_m \exp\left[\sum_q\chi_m^q(b_q{}^\dagger-b_{-q})\right] , \quad (3a)$$

$$B_q = Ub_qU^\dagger = b_q + \sum_m\chi_m^q a_m{}^\dagger a_m . \quad (3b)$$

In the basis generated by these operators, the Hamiltonian (1) with $J_{mn} = 0$ takes the form

$$H = \sum_m \left[E_m - \sum_q |\chi^q|^2 \hbar\omega_q\right]A_m{}^\dagger A_m + \sum_q \hbar\omega_q(B_q{}^\dagger B_q + \tfrac{1}{2}) , \quad (4)$$

from which a two-body term has been neglected since throughout this paper we will only encounter states containing one or zero polarons or excitons.

In developing diagnostics of the dressing process, we will consider two kinds of states: The first, denoted by $|\Phi(t)\rangle$, is that which evolves from an initial state

$|\Phi(0)>$ representing a single bare exciton in a quiescent, undistorted lattice. The second state we shall consider, denoted by $|\Psi(t)>$, is that which evolves from an initial state $|\Psi(0)>$ representing a single polaron in a quiescent, "dressed lattice". Since the bare phonons $(b_q{}^\dagger, b_q)$ and dressed phonons $(B_q{}^\dagger, B_q)$ have different vacuum states when an exciton or polaron is present, we note that

$$|\Phi(0)> = \sum_n \Phi_n(0)\, a_n{}^\dagger(0)\,|0> \quad , \tag{5}$$

$$|\Psi(0)> = \sum_n \Psi(0) A_n{}^\dagger(0)\,|0> \quad , \tag{6}$$

where $|0>$ is the common vacuum annihilated by a_m, b_q, A_n, and B_k and where $\Phi_n(0)$ and $\Psi_n(0)$ are initial amplitudes. The polaron state $|\Psi(t)>$ is stationary, and provides us with the characteristics necessary to recognize the polaron in the non-stationary state $|\Phi(t)>$.

We will not solve the Schrödinger equation for $|\Phi(t)>$ or $|\Psi(t)>$. Rather, we solve the Heisenberg equations for the relevant operators and calculate the required expectation values using the initial states $|\Psi(0)>$ and $|\Phi(0)>$. For the immobile exciton, the Heisenberg solutions are particularly simple, and all subsequent manipulations require only the solutions

$$a_m{}^\dagger(t)a_m(t) = a_m{}^\dagger(0)a_m(0) \quad , \tag{7}$$

$$b_q(t) = \exp[-i\omega_q t]b_q(0) - (1 - \exp[-i\omega_q t])\sum_m \chi_m^q\, a_m{}^\dagger(0)a_m(0) \tag{8}$$

and the relations (3).

Energy

First, let us calculate the expectation value of the total Hamiltonian in the two states:

$$< \Phi(t)\,|H\,|\Phi(t)> = \sum_m E_m P_m + \frac{1}{2}\sum_q \hbar\,\omega_q \quad , \tag{9}$$

$$< \Psi(t)\,|H\,|\Psi(t)> = \sum_m E_m P_m + \frac{1}{2}\sum_q \hbar\,\omega_q - \sum_q |\chi^q|^2 \hbar\,\omega_q \quad , \tag{10}$$

where we have taken $P_m \equiv < \Psi(0)\,|A_m{}^\dagger A_m\,|\Psi(0)> = < \Phi(0)\,|a_m{}^\dagger a_m\,|\Phi(0)>$. The difference in energy between the two states may be interpreted as the polaron binding energy (PBE), and will be taken as the unit of energy in subsequent discussions. Energy is conserved in both states, so if the bare exciton is to evolve into a dressed state containing a polaron, the resulting polaron must coexist with an excess 1 PBE of energy.

Since for our present purposes the exciton is immobile, the only transport of energy which can taken place occurs through the dynamics of the lattice. This inter-system transfer of energy is an essential part of the dressing process, so we consider as a diagnostic the expectation value $< H_{ph}'>$ of the dynamical energy in the normal modes of the isolated lattice; i.e. the total bare phonon energy in excess of the zero-point energy. In the one-polaron state, this energy is time independent and equal to the polaron binding energy.

$$< \Psi(t)\,|H_{ph}'|\Psi(t)> = < \Psi(0)\,|\sum_q \hbar\,\omega_q b_q{}^\dagger(t)b_q(t)\,|\Psi(0)> = 1\ PBE \quad .\tag{12}$$

Since the state $|\Psi(t)>$ contains no dressed phonons at any time, this 1 PBE of energy may be interpreted as the energy stored in the deformation of the lattice which is inherent in the polaron state. On the other hand, the lattice energy in the bare exciton state is time dependent and increases rapidly from zero at the initial time:

$$< \Phi(t)\,|H_{ph}'|\Phi(t)> \equiv < \Phi(0)\,|\sum_q \hbar\,\omega_q b_q{}^\dagger(t)b_q(t)\,|\Phi(0)>$$

$$= K(0) - K(t) \quad . \tag{13}$$

in which we have defined the energy profile function $K(t)$ by

$$K(t) \equiv 2\sum_q |\chi^q|^2 \hbar \omega_q \cos\omega_q t \ . \tag{14}$$

At long times in an infinite system we find that the lattice energy saturates at an asymptotic value equal to twice the polaron binding energy (cf. Fig. 1).

$$\lim_{t \to \infty} <\Phi(t)|H_{ph}'|\Phi(t)> = K(0) = 2 \, PBE \ . \tag{15}$$

Since we have already seen that the quiescent polaron state contains 1 PBE stored in the lattice and a total energy lower than that of the bare exciton by 1 PBE, one is wont to interpret half of the energy (15) as that bound in the polar lattice deformation, and half as the excess which must be radiated if a polaron is to be on formed[20].

The time scale τ_E for this intersystem transfer of energy can be extremely short. In fact, in the elastic limit ($\omega_q \to \upsilon|q|$, where υ is the sound speed) of the one-dimensional problem involving acoustic phonons ($\chi^q \propto |q|^{-\frac{1}{2}}$)[13] $K(t)$ becomes singular and the transfer of energy arbitrarily rapid, so $\tau_E \to 0$. In general, the discreteness of the lattice and the dispersion of the phonons causes τ_E to be finite and the rise of the lattice energy to be non-monotonic.

The energy diagnostic we have considered above are summarized graphically in Figure 1.

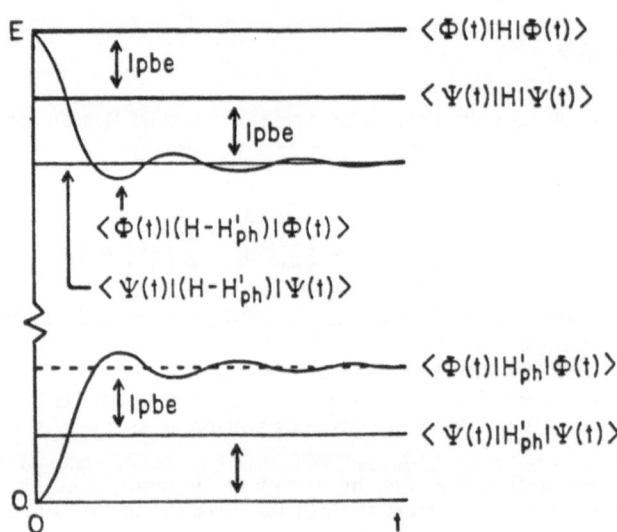

FIGURE 1: Energetics of polaron formation. Expectation values of the total Hamiltonian H, the lattice Hamiltonian (less the zero point energy) H_{ph}, and their difference $H - H_{ph}'$ are exhibited schematically as functions of time. The state vector $|\Phi(t)>$ represents the evolution from an initial state containing a single bare exciton: the state vector $|\Psi(t)>$ represents the evolution from an initial state containing a single polaron.

Lattice Deformation

To further elucidate the dynamics of the intersystem energy transfer considered above, we will calculate the time-dependent displacements of the lattice molecules from their equilibrium positions. This will indicate how molecules both near and far from the excitation region are affected by the dressing processes.

The operator whose expectation value is the displacement of the nth molecule from its equilibrium position is given by

$$u_n(t) = \left(\frac{\hbar}{2NM}\right)^{\frac{1}{2}} \sum_q \frac{\exp[-iqn]}{\omega_q^{\frac{1}{2}}} \{b_q\dagger(t) + b_{-q}(t)\} \tag{16}$$

where M is the mass of the molecular unit comprising a unit cell of the lattice and N is the total number of cells.

In order to determine the shape the lattice deformation in the quiescent polaron state, we use the relations (3) and compute the expectation value $D(n) \equiv \langle \Psi(0) | u_n(t) | \Psi(0) \rangle$:

$$D(n) = \left(\frac{\hbar}{2NM}\right)^{1/2} \sum_q \frac{\exp[-iqn]}{\omega_q^{1/2}} \langle \Psi(0) | \{\exp[i\omega_q t]B_q^{\dagger}(0) \tag{17}$$

$$+ \exp[-i\omega_q t]B_{-q}(0) - 2\sum_m \chi_m^{-q} A_m^{\dagger}(0)A_m(0)\} | \Psi(0) \rangle \quad .$$

Since $| \Psi(0) \rangle$ contains no dressed phonons, the result has the time-independent form

$$D(n) = -\left(\frac{2\hbar}{NM}\right)^{1/2} \sum_{q,m} \frac{\exp[-iqn]\chi_m^{-q}}{\omega_q^{1/2}} P_m \quad . \tag{18}$$

In order to determine the shape of the lattice deformation in the bare exciton state, we similarly compute the expectation value $d(n,t) \equiv \langle \Phi(0) | u_n(t) | \Phi(0) \rangle$:

$$d(n,t) = \left(\frac{\hbar}{2NM}\right)^{1/2} \sum_q \frac{\exp[-iqn]}{\omega_q^{1/2}} \langle \Phi(0) | \{\exp[i\omega_q t]b_q^{\dagger}(0) + \exp[-\omega_q t]b_{-q}(0)$$

$$- 2(1 - \cos\omega_q t) \sum_m \chi_m^{-q} a_m^{\dagger}(0)a_m(0)\} | \Phi(0) \rangle \quad . \tag{19}$$

Since $| \Phi(0) \rangle$ contains no bare phonons we find

$$d(n,t) = -\left(\frac{2\hbar}{NM}\right)^{1/2} \sum_{q,m} \frac{\exp[-iqn]\chi_m^{-q}}{\omega_q^{1/2}} (1 - \cos\omega_q t) P_m \quad . \tag{20}$$

In the elastic limit ($\omega_q \rightarrow v|q|$), this time-dependent lattice deformation can be represented as

$$d(n,t) = D(n) - (\tfrac{1}{2})D(n + vt) - (\tfrac{1}{2})D(n - vt) \quad , \tag{21}$$

where $D(n)$ is the static polaron deformation (17). The dynamics of the lattice can be interpreted straightforwardly as the evolution of the polaron deformation through the radiation of energy. The radiated energy is bound in counterpropagating sound pulses (cf. Fig. 2) having the same shape as the polaron deformation (but the opposite sense; e.g. if the polaron causes the rarefaction of the lattice the radiated energy causes compression). The deformation of the lattice in the bare exciton state evolves on the time scale v^{-1}. This evolution may be much slower than that suggested by the time dependence of the lattice energy whose time scale is τ_E. While this conclusion is reached in the elastic limit, a similar conclusion holds in the presence of dispersion, since the deformation of the lattice involves the radiation of energy, and the propagation of sound energy is limited by maximum group velocity attainable in the given medium. It is interesting to note that in the Einstein oscillator approximation $\omega_q = \omega_0$ for the optical phonons, no persistent deformation develops and there is no net flow of energy into the lattice. In this case, both the energy and deformation of the bare lattice oscillate with period $2\pi/\omega_0$, as is seen by setting $\omega_q = \omega_0$ in (20) and (14). This is consistent with the physical interpretation terms of energy propagation since the phonon group velocity for any q is zero in the Einstein approximation. We stress that these conclusions may be modified in the presence of transport. A nonzero J_{mn} may influence both the shapes of the deformations and the time dependences for the polaron formation and for the radiation of excess energy through the lattice.

Green's Function

Two essential features of the dressing process are thus clearly seen to occur, but on time scales which may be widely separated. It is safe to say that at

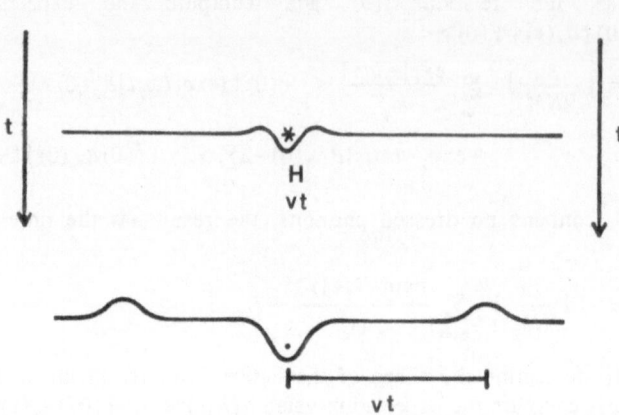

FIGURE 2: Lattice dynamics in polaron formation. The solid line (_____) represents the deformation $d(r,t)$ of the lattice from its equilibrium configuration, a straight line indicating an undistorted lattice. The asterisk (*) indicates the location of the bare exciton created at $t = 0$, and its size indicates the magnitude of the many-body Green's function $- i< \Phi(0) |\Phi(t)>$.

times much greater than υ^{-1} the process is complete and that at times shorter than τ_E it is incomplete, but it is not yet clear on which time scale the identity of the initial bare exciton is lost. The identity of the bare exciton and its time-dependence may be quantified by considering the many-body Green's function[21]

$$G_b(t) = - i< \Phi(0) |\Phi(t)> \tag{22}$$

which measures the projection of the evolving state $|\Phi(t)>$ on the bare exciton state $|\Phi(0)>$. (Note that $|G_b(t)|^2$ is the probability that at time t the state of the many-body system is that corresponding to a bare exciton.) This Green's function can be calculated exactly using the canonical transformation (2). For simplicity we consider a bare exciton initially localized in a single site-state r. for such an initial state

$$G_b(t) = - i< 0 |a_r(t)a_r^\dagger(0) |0>$$
$$= - i< 0 |A_r(t)\exp[-S_r(t)]\exp[S_r(0)]a_r^\dagger(0) |0> \quad , \tag{23}$$

where we have used the relations (3) and defined

$$S_r(t) \equiv \sum_q \chi_r^q \{\exp[i\omega_q t]B_q^\dagger(0) - \exp[-i\omega_q t]B_{-q}(0)\} \quad . \tag{24}$$

Since A_r commutes with H its time dependence may be extracted explicitly $A_r(t) = - \exp[i\Omega_r t]A_r(0)$, where $\hbar\Omega_r = < \Psi |H |\Psi>$), allowing the operator product to be expressed entirely in terms of initial values. The operators $A_r^\dagger(0)$ and $A_r(0)$ commute with $\exp[S_r(t)]$, which allows $G_b(t)$ to be reduced to[20]

$$G_b(t) = - i< 0 |\exp[-S_r(t)]\exp[S_r(0)] |0> \exp[(-i\Omega_r t)] \quad . \tag{25}$$

Recognizing $\exp[S_r(t)]$ as a product of coherent state displacement operators, the well-known coherent state inner product[22]

$$< \alpha | \beta > \; = \exp[\alpha^* \beta - (\tfrac{1}{2}) |\alpha|^2 - (\tfrac{1}{2}) |\beta|^2] \qquad (26)$$

may be used to obtain the desired result

$$G_b(t) = -i\exp\{- \sum_a |\chi^q|^2 [1 - \exp(-i\omega_q t)]\}\exp[-i\Omega_r t] \; . \qquad (27)$$

Comparing this result with the lattice deformation result (20) with $P_m = \delta_{mr}$, we see that the real part of the exponent in (27) is proportional to $d(r,t)$, the lattice deformation at the occupied site, when the coupling coefficient appropriate to acoustic phonons is used ($\chi^q \propto |q|^{-\frac{1}{2}}$). Thus, the time dependence controlling the departure of $|\Phi(t)>$ from the initial bare exciton state is seen to be that of sound propagation rather than that of intersystem energy transfer.

The projection of $|\Phi(t)>$ on the one-polaron state may be similarly calculated. For comparison we define

$$G_p(t) = -i< \Psi(0) | \Phi(t)>$$

$$= -i\exp[-\tfrac{1}{2} \sum_q |\chi^q|^2]\exp(-i\Omega_r t) \; . \qquad (28)$$

The first notable feature of this result is the constancy of its magnitude in time. While it may seem counterintuitive that the magnitude of the projection of the evolving many-body state on the one-polaron state should be time independent (which might yield the impression that the bare exciton does not "dress"), this is just a consequence of the fact that the one-polaron state we have considered is an eigenstate of our one-body Hamiltonian: the only time dependence exhibited by a projection onto a Hamiltonian eigenstate is a precession of phase at the eigenfrequency of the eigenstate.

The second notable feature of this result is its relation to the long-time limit of $G_b(t)$. We find

$$|G_p(\infty)| > |G_b(\infty)| = |G_p(\infty)|^2 \; . \qquad (30)$$

That is, at long times the overlap $iG_b(\infty)$ of $|\Phi(t)>$ with the bare exciton state is smaller in magnitude than the overlap $iG_p(\infty)$ of the same many-body state with the polaron state, regardless of coupling strength. Thus, in the sense measured by inner products, the final state of the system is more similar to a polaron state than $\hat{a}$ to bare exciton state.

RELEVANCE TO TRANSPORT

The foregoing analysis has been carried out in the $J_{mn} = 0$ limit of the standard model of exciton transport: that is, for immobile excitons coupled linearly to phonons. The relevance of our conclusions to the dynamics of mobile excitons is obviously open to question. It is safe to say that for sufficiently narrow exciton bands, the response of the lattice to initial excitation should be similar to that seen above both qualitatively and quantitatively. However, even in this case, no direct information about exciton transport is provided by these results without reformulating the evolution equations for the basic operators. The reformulation leads to the exciton transport theory recently advanced by West and Lindenberg[17].

The theory of West and Lindenberg exploits the fact that for the given Hamiltonian the Heisenberg equations of motion for the phonon operators (e.g., b_q) can be integrated to yield solutions for these operators which are simple functionals of the exciton operators $a_m^\dagger, a_m$:

$$b_q(t) = \exp[-i\omega_q t]b_q(0) - i\int_0^t d\tau \exp[-i\omega_q(t-\tau)]\sum_n \chi_n^q \hbar \omega_q a_n^\dagger(\tau)a_n(\tau) \; . \quad (41)$$

Just as (8) can be understood as the response of the lattice to the disturbance caused by the creation of an immobile exciton at the initial time, (41) may be understood as the continuation of this response to continuing disturbances caused by the motion of the exciton. Using (41) and its hermitian conjugate to eliminate the phonon operators from the Heisenberg equation for $a_r(t)$ yields the essential equation of the theory

$$a_r(t) = \sum_n J_{nr} a_n(t) + \sum_n [K_{rn}(0) a_n^\dagger(t) a_n(t) - K_{rn}(t) a_n^\dagger(0)] a_r(t)$$

$$+ F_r(t) a_r(t) + \int_0^d d\tau \sum_n K_{rn}(t-\tau) \frac{\partial}{\partial \tau} [a_n^\dagger(\tau) a_n(\tau)] a_r(t) \tag{42}$$

in which the integral appearing in (41) has been integrated by parts. The fluctuating operator $F_r(t)$ and kernel $K_{rn}(t)$ are respectively defined by

$$F_r(t) = \sum_q \chi_r^q \, \hbar \, \omega_q \, [\exp(i \omega_q t) b_q^\dagger(0) + \exp(-i \omega_q t) b_{-q}(0)] \tag{43}$$

and

$$K_{rn}(t) = 2 \sum_q \chi_r^q \chi_n^{-q} \, \hbar \, \omega_q \cos(\omega_q t) \; . \tag{44}$$

The existence of a fluctuation-dissipation relation between the kernels and fluctuating operators is one of the compelling motivations for the approach.

Comparison of the surface terms in (42) with the form of the lattice deformation $d(r,t)$ in (20) suggests the interpretation of $-K_{rn}(t) a_n^\dagger(0) a_n(0)$ as the potential due to the initial counter-propagating pair of lattice distortions, and of $K_{rn}(0) a_n^\dagger(0) a_n(t)$ as the potential due to a polaronic lattice deformation bound to the mobile exciton. The integration in (42) may then be interpreted as contributing time-dependent corrections to these potentials. Since the corrections involve the propagation of disturbances through the lattice at finite velocities, this term is inherently non-local in both space and time.

An important relation between the mobile and immobile exciton evolutions comes through the form of the kernel $K_{rn}(t)$. Regardless of dispersion, we find on comparing (44) and the immobile exciton results (13)-(15) that

$$K_{rr}(t) = K(t) = 2 \, PBE - \langle \Phi(t) | H_{ph}' | \Phi(t) \rangle \; . \tag{45}$$

Thus, both the temporal structure and physical origins of the dissipative kernel are found to be identical to that of the fundamental dissipative process, *i.e.* the exchange of energy between the degrees of freedom regarded as system and bath.

CONCLUSION

We have considered the standard model for exciton transport in the immobile exciton limit in which the model is exactly solvable. In this limit we have analyzed the evolution of the many-body system from initial states representing: 1) a bare exciton in a quiescent bare lattice, and 2) a polaron in a quiescent dressed lattice. As diagnostics of the many-body state evolution, we have calculated exactly: a) the total energy, b) the dynamical energy content of the bare lattice, c) the deformation of the bare lattice, and d) many-body Green's functions.

The picture of the dressing process which emerges from this analysis involves two energy transfer processes which may occur on different time scales. Neither process is that usually connoted by the term "energy transfer", since this term is usually applied to transport of energy through the resonant processes we have neglected in assigning J_{mn} the value zero. The first energy transfer process, schematized in Figure 1, is the transfer of energy from the bare exciton space into the bare lattice space. We have referred to this process as intersystem energy transfer, since in the present context there can be no confusion with singlet-triplet conversion. The second energy transfer process, depicted in Figure 2, is the dispersal of the energy transferred into the bare lattice space. We have referred to this process as sound propagation, since we have been able to identify the sound speed as the parameter controlling the dispersal.

We have been able to conclude that while intersystem energy transfer occurs on a time scale which may be arbitrarily short, the dressing process by which a bare exciton evolves toward a polaron-like state proceeds on the time scale of sound propagation. This conclusion is reached on the basis of the appearance of similar time dependencies in the approaches of the bare lattice deformation and many-body Green's functions to their asymptotic forms, while a distinct time dependence is found for the intersystem energy transfer.

The bare lattice in its asymptotic state is seen to contain a stationary distortion, identical to that found for the polaron, together with paired distortions propagating at the sound velocity away from the excitation region. Green's functions show that the asymptotic many-body state is more similar to a polaron state than to a bare exciton state; however, the polaron "component" of the wave function maintains a constant magnitude throughout the dressing process. The authors have shown elsewhere that the reduced density matrix representing an ensemble of initially bare excitons is found at long times to approach the density matrix representing an ensemble of a similarly prepared polarons[19]. Together, these results show that while the asymptotic state of an initially bare exciton is only polaron- like. The qualifier is needed only to specify a condition of the lattice which is irrelevant to the reduced description of the asymptotic state.

The relevance of the immobile exciton dynamics to exciton transport is established through the identification of the dissipative kernel in the exciton transport theory of West and Lindenberg with the time-dependent profile of the bare lattice energy in the immobile exciton model.

ACKNOWLEDGEMENT

This work was supported in part by NSF DMR-8315950, by DARPA DAAG 29-85-K-0246 and by the Independent Research Funds of the La Jolla Institute.

REFERENCES

1. V. M. Kenkre, Exciton Dynamics in Molecular Crystals and Aggregates, Springer Tracts in Modern Physics, Vol. 94 (Springer Verlag, Berlin, 1982).
2. P. Reineker, Exciton Dynamics in Molecular Crystals and Aggregates, Springer Tracts in Modern Physics, Vol. 94 (Springer Verlag, Berlin, 1982).
3. A. S. Davydov, Phys. Stat. Sol. 30, 357 (1963); Physica Scripta 20, 387 (1979); Sov. JETP 51, 397 (1980).
4. A. C. Scott, Phys. Rev. A26, 578 (1982); P. S. Lomdahl and W. C. Kerr, Phys. Rev. Lett. 55, 1235 (1985).
5. G. Careri, U. Buontempo, F. Galluzzi, A. C. Scott, E. Gratton, and E. Shymsunder, Phys. Rev. B30, 4689 (1984).
6. A. S. Davydov, Sov. Phys. Usp. 25, 898 (1982).
7. G. S. Edwards, C. C. Davis, J. D. Saffer, and M. L. Swicord, Phys. Rev. Lett. 53, 1284 (1984).
8. A. C. Scott, Phys. Rev. A31, 3518 (1985).
9. S. J. Webb, Phys. Rev. A31, 3518 (1985).
10. E. Balanovski and P. Beaconsfield, Phys. Rev. A32, 3059 (1985).
11. H. Fröhlich, Proc. R. Soc. London, Ser. A215, 291 (1952); Adv. Phys. 3, 325 (1954).
12. T. Holstein, Ann. Phys. (NY) 8, 325 (1959); 343 (1959).
13. J. Schnakenberg, Phys. Stat. Sol. 28, 623 (1968).
14. V. M. Kenkre, Phys. Rev. B11, 1741 (1975); 12, 2150 (1975).
15. R. Silbey, Ann. Rev. Phys. Chem. 27, 203 (1976).
16. D. R. Yarkony and R. Silbey, J. Chem. Phys. 65, 1042 (1976); 67, 5818 (1977).
17. B. J. West and K. Lindenberg, J. Chem. Phys. 83, 4118 (1985).
18. B. W. Brown, K. Lindenberg, and B. J. West, J. Chem. Phys. 83, 4136 (1985).
19. D. W. Brown, K. Lindenberg, and Bruce J. West, J. Chem. Phys. 84, 1574 (1986).
20. S. Nakajima, Y. Toyozawa, and R. Abe in The Physics of Elementary Excitations, Springer Series in Solid-State Sciences, Vol. 12, eds. M. Cardona, P. Fulde and H. -J. Queisser (Springer Verlag, New York, 1980).
21. E. N. Economou, in Green's Functions in Quantum Physics, Springer Series in Solid-State Sciences, Vol. 7, eds. M. Cardona, P. Fulde and H. -J. Queisser (Springer Verlag, New York, 1983).
22. R. J. Glauber, Phys. Rev. 131, 2766 (1963).

EXCITRON TRANSPORT IN MIXED CRYSTALS: QUANTUM PERCOLATION

L. Root and J.L. Skinner

Department of Chemistry
Columbia University
New York, NY 10027

Two-component mixed crystals are interesting and important systems for studying exciton transport. At high temperatures, a hopping description of exciton dynamics is thought to be adequate.[1] Considering excitations on only one species of molecule, and assuming that the hopping rates for these active sites are nonzero only for nearest neighbor pairs, this then defines a classical percolation problem.[2] That is, if the fraction of active sites is p, then for p less than a critical value p_c, only finite clusters of active molecules exist and no long-range transport is possible. However, if $p > p_c$, an infinite cluster exists, and long-range transport occurs. It is known that for a 2-d square lattice $p_c \cong 0.59$, while for a 3-d simple cubic lattice $p_c \cong 0.31$.

At low temperatures, a classical hopping description of transport is inadequate, and it is customary to describe the active molecules by a quantum mechanical tight-binding Hamiltonian.[1] If the transfer matrix elements are nonzero only for nearest neighbors, this defines a quantum percolation problem. It is clear that for $p < p_c$ there can be no long-range transport because there is no infinite cluster of active sites. On the other hand, for p = 1, the system is translationally invariant and Bloch's theorem states that the eigenstates are delocalized plane waves, implying the existence of long-range transport. Thus it is natural to ask whether there exists a critical fraction, p_q, with $p_c \leq p_q \leq 1$, at which transport occurs.

The quantum percolation transition is intimately connected with the nature of the eigenstates: below p_q all states are localized, while above

p_q at least some states are extended. Thus the quantum percolation problem is really an Anderson localization problem.[3] Recently, a scaling argument has been proposed which states that for an infinitesimal amount of disorder, for $d \leq 2$, all eigenstates are localized.[4]

We have performed a real-space renormalization group calculation of the quantum percolation problem. Our method is based on phenomenological[5] or macroscopic[6] renormalization, in which the correlation lengths of different sized systems are related in order to obtain approximations to the critical point, p_q. Our preliminary results give $p_q \cong 0.45$ for a simple cubic lattice.[7] This is to be compared with recent calculations of p_q by Raghavan,[8] Srivastava and Chaturvedi,[9] and Odagaki and Chang[10] giving 0.38, 0.47, and 0.70, respectively, and with the classical value of $p_c \cong 0.31$. Thus it appears that indeed $p_q > p_c$. Our preliminary result for a square lattice is $p_q = 1$, in agreement with the scaling prediction.[4]

ACKNOWLEDGMENTS

We thank the National Science Foundation for support from grant No. DMR 83-06429. L.R. acknowledges additional support in the form of an NSF Graduate Fellowship. J.L.S. acknowledges additional support in the form of a Camille and Henry Dreyfus Teacher-Scholar Award, an Alfred P. Sloan Fellowship, and an NSF Presidential Young Investigator Award.

REFERENCES

1. R. Silbey, in Spectroscopy and Excitation Dynamics in Condensed Molecular Systems, edited by V.M. Agranovich and R.M. Hochstrasser (North Holland, Amsterdam, 1983).
2. D. Stauffer, Phys. Rep. 54, 1 (1979).
3. P.W. Anderson, Phys. Rev. 109, 1492 (1958).
4. E. Abrahams, P.W. Anderson, D.C. Liciardello, and T.V. Ramakrishnan, Phys. Rev. Lett. 42, 673 (1979).
5. M.P. Nightingale, Physica 83A, 561 (1976).
6. W.L. McMillan, Phys. Rev. B 31, 344 (1985).
7. L. Root and J.L. Skinner, to be published.
8. R. Raghavan, Phys. Rev. B 29, 748 (1984).
9. V. Srivastava and M. Chaturvedi, Phys. Rev. B 30, 2238 (1984).
10. T. Odagaki and K.C. Chang, Phys. Rev. B 30, 1612 (1984).

ON ELECTRON TRANSFER REACTIONS IN DISORDERED MEDIA

E. Canel

The Rockefeller University
1230 York Avenue
New York, NY 10021

It is the purpose of this paper to develop unified mathematical models
for electron transfer reactions in condensed media. Using these models we
will discuss the main qualitative features of these processes. Electron
transfer reactions in the gas phase are reasonably well understood. In some
cases the reaction is direct; in others it involves the formation of an unstable quasimolecule (of lifetime 10^{-13}sec) posessing a full set of quantum
numbers.[1] [2] [3] [4] Using standard "adiabatic techniques" the processes can
be described theoretically.[5][6] In condensed media the situation is complicated by various factors such as:

(1) Electrostatic interactions with the solvent that modify the energy
     levels and wave functions of the reactants.
(2) Polarization changes in the solvent.
(3) Bridge formation by solvent molecules·
(4) Relaxation processes which do not occur in the gas phase.
(5) The symmetry of interacting orbitals·

To construct models which describe the various aspects of electron
transfer in a unified way it is convenient to formulate the problem using
Green Functions. In the following, such a method will be developed. For
the sake of clarity we consider a simple model consisting of donor, acceptor
and one electron. We will assume that the positions of donor and acceptor
can be characterized by the vectors $R_D(t)$ and $R_A(t)$ respectively the Schrodinger
equation for the electron is then

$$(1) \qquad \frac{\hbar}{i} \frac{\partial \psi}{\partial t} = \left[\frac{p^2}{2m} + V_D\big(r - R_D(t)\big) + V_A\big(r - R_A(t)\big) \right] \psi$$

with the initial condition that the electron is bound to the donor. To
solve this problem we assume that only a few donor and acceptor orbitals are
important for the reaction. For example, in the simplest two level case we
set

$$(2) \qquad \psi = a(t)\psi_A\big(r - R_A(t)\big) + d(t)\psi_D\big(r - R_D(t)\big)$$

where ψ_D and ψ_A are the wave functions of the isolated donor and acceptor.
From (2) we find

(3)
$$\frac{\hbar}{i}\frac{\partial}{\partial t}\begin{pmatrix} a \\ d \end{pmatrix} = \begin{pmatrix} H_{aa} & H_{ad} \\ H_{da} & H_{dd} \end{pmatrix}\begin{pmatrix} a \\ d \end{pmatrix}$$

where initially d=1 and a=0 and $H_{ad} = \langle \psi_A | \frac{p^2}{2m} + V_D + V_A | \psi_D \rangle$ etc.
In general we would obtain

(4)
$$\frac{\hbar}{i}\frac{\partial X}{\partial t} = HX$$

where H is the appropiate matrix and initally $X = X_0$ The matrix H
formally identical to the matrix found in the usual LCAOMO approximation.
In the present case, however, the matrix elements are in general time
dependent.   The initial value problem (4) is solved by introducing a
Green Function $\mathcal{G}$ defined by

(5)
$$\left[\frac{\hbar}{i}\frac{\partial}{\partial t} - H\right]\mathcal{G} = I\delta(t - t_1)$$

The matrix elements of $\mathcal{G}$ are the transition amplitudes of interest; in
our two level case for example the electron transfer amplitude is $\mathcal{G}_{ad}$
If $R_A(t)$ and $R_D(t)$ vary slowly with time the approximate solution of (5) is

(6)
$$\mathcal{G}(t,t_1) = \sum_{\alpha} P_{\alpha}(t) \exp\left\{-i\int_{t_1}^{t} E_{\alpha}(\tau)\,d\tau\right\}\Theta(t - t_1)$$

where $E_{\alpha}(t)$ are the eigen values of H and P_{α} is the corresponding projection operator.   Using (6) we can calculate the charge transfer probability

For the two level case we find

(7)
$$|\mathcal{G}_{ad}(t - t_1)|^2 = \frac{1}{2}\sin^2\left\{\int_{t_1}^{t} H_{ad}(\tau)\frac{d\tau}{\hbar}\right\}$$ if $E_A = E_D$

and

(8)
$$|\mathcal{G}_{ad}(t - t_1)|^2 \sim \left|\frac{H_{ad}(t - t_1)}{E_A - E_D}\right|^2$$ if $E_A \neq E_D$

Since in general $H_{ad}(t) \to 0$ as $t \to \infty$ there will be no electron transfer
in the second case.   Since the model is non dissipative this result simply
expresses energy conservation.   More complicated systems (involving more
levels) can be treated in the same way.   To conclude the discussion of the
method it should be noted that the Green Function method can also be used
to calculate transition rates.   To see this we write

(9)
$$H = H_0 + \Gamma$$

where H_0 is the diagonal part of H . Define a
transition matrix T by

(10a)
$$T = \Gamma + \Gamma\mathcal{G}\Gamma$$

the transition rate is then

(10b)
$$\frac{|T_{ad}|^2}{\hbar}\delta(E_A - E_D)$$

where E_D and E_A are the appropiate initial and
final states.

118

The previous discussion was based on a LCAO approximation for H. It can however be refined by using more elaborate molecular orbital methods. The essential assumptions made are

(1) Slow motion of donor and acceptor
(2) Well defined initial and final states
(3) Expansion of wave functions in terms of a finite number of frontier orbitals

In the following we will try to extend the method to discuss electron transfer reactions in condensed media. The fundamental theory of these processes is due to Marcus, Levitch, Dorgonadze and Hush MLDH.[7] Recently Calef and Wolyness[8] have extended the theory using Kramers diffusion theory.

According to these authors the expression for the rate can be written as the product of two factors, one representing the "tunneling" probability of the electron, the other, a temperature dependent term connected to the activation energy of the process. In the following we will discuss both factors. We will begin with the electronic term, which will dominate if the activation energies are small. In such cases the transfer rate depends upon

(1) The position and width of relevant donor and acceptor levels
(2) Bridges (or barriers) formed by the solvent molecules
(3) The overlap between relevant wave functions
(4) Relaxation processes connected with the solvent

We will discuss these in turn. Electron transfer can occur only if donor and acceptor levels have the same energy. In condensed matter the energy levels of the reactants will be broadened and shifted and the interactions with the collective modes will provide new paths for charge transfer. To study these effects we will initially assume that the transfer system can be described by the Schrodinger equation.

$$(11) \qquad \frac{\hbar}{i}\frac{\partial \psi}{\partial t} = \left(\frac{p^2}{2m} + V_D + V_A + V_{Sol} \right) \psi$$

where V_{sol} is the effective potential due to the solvent. Since the medium is disordered V_{sol} will be assumed to be a random potential.

Because of this we introduce a Green Function formulation which combines the molecular orbital theory discussed above with the theory of disordered systems. The equation for the one particle Green Function g in the medium is defined by

$$(12) \qquad \left[\frac{\hbar}{i}\frac{\partial}{\partial t} - \left(\frac{p^2}{2m} + V_D + V_A + V_{Sol} \right) \right] g(t - t_1, x - x_1) = \delta^4(t - t_1, x - x_1)$$

If we assume that the wave function   can be approximated by a suitable linear combination of donor and acceptor wave function, we can contruct a matrix $\mathcal{G}$ defined by

$$(13) \qquad \left[\frac{\hbar}{i}\frac{\partial}{\partial t} - \tilde{H} \right] \tilde{\mathcal{G}} = I\delta(t - t_1)$$

where   for a two level case

$$\tilde{H} = \begin{pmatrix} \tilde{E}_A & \tilde{\Gamma} \\ \tilde{\Gamma} & \tilde{E}_D \end{pmatrix}$$

here $\tilde{E}_{A,D} = \int \psi_{A,D}^* \left(\frac{p^2}{2m} + V_D + V_A + V_{Sol} \right) \psi_{A,D} \, dVol$

and $\tilde{\Gamma} = \int \psi_A^* \left(\frac{p^2}{2m} + V_D + V_A + V_{Sol} \right) \psi_D \, dVol$

Clearly $\tilde{E}_A$ and $\tilde{E}_D$ random variables.

Within the tight binding approximation, we assume that the broadened donor and acceptor levels $\tilde{E}_D$ and $\tilde{E}_A$ can be characterized by distribution functions $\rho(\tilde{E}_A)$ and $\rho(\tilde{E}_D)$ defined by

where
$$\rho(\tilde{E}_{A,D}) = \int \operatorname{Im} g_{a,d}(x,x,E) \, d^3x$$

(14)
$$\left[E - \frac{p^2}{2m} + V_{A,D} + V_{Sol} \right] g_{a,d}(x,x_1,E) = \delta^3(x - x_1)$$

The calculation of the overlap Γ depends upon the knowledge of the wave functions in the medium and will be discussed in detail later. Once ρ_A and ρ_D are known the transfer probability and transfer rates can easily be calculated. Thus we find that the probability for charge transfer is

(15)
$$|g_{ad}(\infty, -\infty)|^2 = \int d\tilde{E}_A \, d\tilde{E}_D \, \rho(\tilde{E}_A)\rho(\tilde{E}_D)\delta(E_A - E_D) \sin^2 \left\{ \int_{-\infty}^{\infty} \Gamma \frac{dt}{\hbar} \right\}$$

while first order perturbation theory gives

(16)
$$\int d\tilde{E}_A \, d\tilde{E}_D \, \frac{\rho(\tilde{E}_A)\rho(\tilde{E}_D)}{\hbar} \Gamma^2(E_{A,D}, R_{A,D})\delta(E_A - E_D)$$

for the instantaneous transfer rate (16) is meaningful only if the change transfer is very fast compared to the motion of donor and acceptor.

Qualitatively two facts are important. In the first place the broadening of the levels will in general assist electron transfer. In the second place the coupling integral Γ will be renormalized and become frequency dependent. Broadly speaking the magnitude of Γ depends on whether the medium is localizing or not. To discuss this the coupling mechanism between donor and acceptor states must be considered in more detail. Within our mode two limiting coupling cases are possible. Either the solvent molecules can act as a bridge (through bond coupling) or the coupling's direct (through space coupling). In the following we discuss both cases.

A)   Through Bond Coupling

The through bond coupling mechanism was firt proposed by Hoffmann[9] to explain the splitting of levels associated with lone pair orbitals of certain molecules. For the cases he considers Hoffmann shows that through space interaction between lone pair orbitals is too small to explain the observed splitting. To do this the interaction with the σ and σ^* orbitals associated with the intervening C–C bonds must be considered. Interactions of this type are also important for the electron transfer problem[10]. The question is which solvent states should be included in the coupling. To study this problem we consider a model consisting of one electron, one donor and a bridge. The equation for the Green Function is then

$$(17) \qquad \frac{\hbar}{i}\frac{\partial \mathcal{G}}{\partial t} = H\mathcal{G} + I\delta(t - t_1); \qquad H = \begin{pmatrix} E_A & \Gamma & 0 \\ \Gamma & E_B & \Gamma \\ 0 & \Gamma & E_D \end{pmatrix}$$

For the simple case $E_A = E_D \ll E_B$ we find

$$(18) \qquad |\mathcal{G}_{a,d}(\infty, -\infty)|^2 = \frac{1}{2}\sin^2\left\{\int_{-\infty}^{+\infty} \frac{\Gamma^2(t)\,dt}{E_B - E_A}\right\}$$

and

$$(19) \qquad |T_{da}|^2 = \left|\frac{\Gamma^2}{E_B - E_A}\right|^2 \delta(E_A - E_D)$$

The "energy levels" of $H(t)$ (in the same approximation) are given by

$$E_A - \frac{\Gamma^2}{E_B - E_A} \qquad E_A \qquad E_A + \frac{\Gamma^2}{E_B - E_A} \qquad E_D = E_A$$

A direct connection between transfer rate and level splitting is evident. This result however, only holds for a one electron system as the following model shows. Consider a four site system whose energies are indicated in Figure 1.

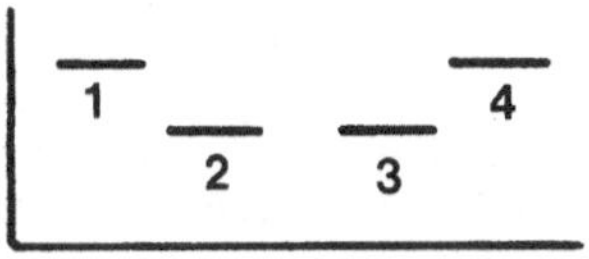

Fig. 1

Assuming nearest neighbor coupling the Hamiltonian can be written in the form

$$(20) \qquad H = \sum_{i=1}^{4} \epsilon_i A_{i\sigma}^\dagger A_{i\sigma} + \sum \Gamma(A_{i\sigma}^\dagger A_{i+1\sigma} + \text{c.c.})$$

where $A_{i\sigma}^\dagger$ and $A_{i\sigma}$ are the usual Fermion creation and destruction operators and σ denotes the electron spin.

H can be diagonalized by the canonical transformation U and we find

$$(21) \qquad \tilde{H} = \sum E_i \tilde{A}_{i\sigma}^\dagger \tilde{A}_{i\sigma}$$

$$\tilde{A}_{i\sigma} = \sum_{j=1}^{4} U_{ij} A_{j\sigma}$$
$$A_{i\sigma} = \sum_{j=1}^{4} U_{ij}^{-1} \tilde{A}_{j\sigma}$$

if the system contains only one electron, the transition amplitude 4 to 1, $\mathcal{G}_{14}$ is given by

(22) $\mathcal{G}_{14} = \langle 0|e^{iHt/\hbar}A_{1\sigma}e^{-iHt/\hbar}A_{4\sigma}^{\dagger}|0\rangle = -\sum_{j=1}^{4}e^{-iE_jt/\hbar}U_{1j}U_{4j}^{*}$

If on the other hand the system contains two additional electrons in the groundstate we have

(23) $\mathcal{G}_{14} = \langle \phi|e^{i\tilde{H}t/\hbar}A_{1\sigma}e^{-i\tilde{H}t/\hbar}A_{4\sigma}^{\dagger}|\phi\rangle$

$|\phi\rangle = \tilde{A}_{1\sigma}^{\dagger}\tilde{A}_{2\sigma}^{\dagger}|0\rangle$

and we find

(24)

$$\mathcal{G} = \sum_{k}U_{1k}U_{4k}^{*}e^{iE_kt/\hbar}(-\delta_{kj} + \delta_{j1}\delta_{k1})$$

$$= \sum_{k=2}^{4}U_{1k}U_{4k}^{*}e^{-iE_kt/\hbar}$$

Transitions to the ground state are thus excluded as required by the exclusion principle. On the other hand all levels must be included in the calculation of engery levels (and the associated splitting.) For the σ bond model considered, this means that both σ and σ^{*} orbitals must be included in the calculation of stationary states, while only σ orbitals occur in the dynamic problem. This conclusion is however, based on the assumption that donor and acceptor states are localized and that a tight binding approximation is reasonable.

It has been assumed that electron-electron interactions and excited molecular states are not important. In particular excited states must be considered in the case of light induced electron transfer reactions in which electron-electron interactions involve exciton formation. As a result new channels for charge transfer may become available.

For example consider the Hamiltonian

(25) $H = \sum_{i=1}^{N}\epsilon_i A_{i\sigma}^{\dagger}A_{i\sigma} + \sum(A_{i\sigma}^{\dagger}A_{i+1\sigma} + \text{c.c.})$

and assume the ground state to be $\prod_{i<f}A_i^{\dagger}|0\rangle$

Define hole creation and destruction operators by

(26) $B_i = A_i^{\dagger}$ if $i > f$

$B_i^{\dagger} = A_i$ if $i \ll f$

H then becomes

(27)

$$H = \sum_{i>f} \epsilon_i^e A_{i\sigma}^\dagger A_{i\sigma} + \sum_i \left[\Gamma A_{i\sigma}^\dagger A_{i+1\sigma} + \text{c.c.} \right] - \left[\sum_{i\leq f} \epsilon_i^h B_{i\sigma}^\dagger B_{i\sigma} + \Gamma \sum B_{i\sigma}^\dagger B_{i+1\sigma} + \text{c.c.} \right]$$

The one electron Hamiltonian has thus been replaced by an equivalent operator describing electrons and holes; change transfer is possible by either charge carrier.

The interaction of electrons and holes can lead to exciton formation and thus electron transfer can occur by exciton transfer from donor to acceptor followed by hole transfer in the opposite direction. Since the coupling parameters for these processes differ, the rates of electron and hole transfer will differ as well.

In particular it is possible to construct systems which (due to symmetry of the orbitals involved) will favour one or the other mechanism.

If the appropiate states for through bond coupling do not exist charge transfer will occur through space. We discuss this below.

B) Through Space Coupling

The magnitude of this coupling depends upon the range of the localized donor and acceptor wave functions, the energy levels of the states of interest and the effective potentials in the medium.

To discuss these points we will consider a model consisting of an electron bound to an impurity immersed in a solvent formed by a disordered assembly of non-polar molecules.

The effect of the medium on the electronic states of the medium is twofold. Because of the exclusion principle the electronic impurity states must be orthogonal to the "core states" of the solvent. In addition electron-electron interactions will screen the impurity potential. It is convenient to discuss the orthogonality problem first.

Assuming that the solvent core states are known we can deal with the orthogonality problem by introducing a pseudo potential similar to the one used in the study of crystaline solids.[11] The one electron Green Function is then defined by

(28a)

$$\left(E - \frac{p^2}{2m} \right) g(x, x_1, E) + \sum_{\alpha, R_n} \int d^3 x_2\, \psi_\alpha^*(x - R_n)\psi_\alpha(x_2 - R_n) g(x_2, x_1, E)(E - E_\alpha)$$

$$= \delta^3(x - x_1)$$

or

$$g(x, x_1, E) = g_0(x, x_1, E)$$

$$\text{(28b)} \qquad + \sum_{\alpha, R_n} \int d^3x_2 d^3x_3 \, g_0(x, x_2, E) \psi_\alpha^*(x_2 - R_n)$$

$$\times \psi_\alpha(x_3 - R_n)(E - E_\alpha) g(x_3, x_1, E)$$

here the random variable R_n denotes the position of the solvent molecules, α the quantum numbers of the core states and E_α the core state energy.

Since we are dealing with localized states we are interested in $E < 0$. In the following we will study the properties of g which will be obtained by first solving 28a) by interation and then averaging the resulting series term by term over R_n. The procedure is familiar from the theory of disordered solids. The series for g can be represented by diagrams in the usual way

we find

an approximate expression for g is then obtained by resumming an appropiate subset of diagrams. We note that the density dependence of a diagram is indicated by the number of crosses it contains.

The summ of the leading terms in the high density limit can thus be obtained by solving the integral equation

The analytic form of this equation in momentum space is

(29)

$$g_{av}(k, k_1, E) = g_0(k, E)\delta(k - k_1)$$

$$+ \sum_\alpha (E - E_\alpha) \int d^3l \, g_0(k, E) \psi_\alpha(k) \psi_\alpha(l) \rho(k - l) g_{av}(l, k_1, E)$$

$$+ \int d^3q_1 d^3q_2 d^3q_3 \sum_{\alpha_1, \alpha_2} g_0(k, E)(E - E_{\alpha_1})(E - E_{\alpha_2}) \psi_{\alpha_1}^*(k) \psi_{\alpha_1}(q_1)$$

$$\times g_{av}(q_1, q_2, E) \psi_{\alpha_2}^*(q_2) \psi_{\alpha_2}(q_3) \rho(q_1 - q_2 - k + q_3) g_{av}(q_3, k_1, E)$$

where $\rho(k)$ is the Fourier transform of the pair distribution function of the solvent molecules (with respect to the solute). The qualitative properties of g_{av} can most easily be studied for the case of total disorder (S=constant)

we then find

(30)

$$g_{av}(k, E) = g_0(k, E)\delta(k - k_1) + \sum_\alpha \rho d^3l \, g_0(k, E) |\psi_\alpha(l)|^2 (E - E_\alpha) g_{av}(k, E)$$

$$+ \sum_{\alpha, \alpha_1} \int d^3l \, g_0(k, E) |\psi_\alpha(k)|^2 |\psi_{\alpha_1}(l)|^2 g_{av}(l, E) g_{av}(k, E)(E - E_\alpha)(E - E_{\alpha_1})$$

124

In the simple case of one core level the approximate form of $g_{av}(x, x_1, E)$ is coordinate space can easily be found if $\left|\frac{E - E_\alpha}{E}\right| \ll 1$

we obtain

(31)
$$g_{av}(x, x_1, E) = \frac{\exp\left\{-\sqrt{\frac{2m}{\hbar}\left[|E| + \Omega\rho(E - E_\alpha)\right]}\left|x - x_1\right|\right\}}{|x - x_1|}$$

where ρ is the concentration of solvent molecules Ω their molecular volume qualitatively the medium acts as a repulsive barrier of order $\Omega\rho(E - E_\alpha)$ thus tending to localize the electrons bound to the impurity.

To study this further we next examine the effective potential of an ion in the same approximation.

If an impurity is added to the solvent the formal diagramatic expansion of the Green Function describing bound states takes the form

were ● represents the impurity potential

The simplest equation describing a bound state is

were ▭ → g_{AV} and ⊙ is the renormalized potential. In the simplest case (ρ = constant)

or

(32) $V_{eff}(k, k_1, E) = V^0(k, k_1) + \rho \sum_{\alpha, \alpha_1} \int d^3l\, \psi_\alpha^*(k)\psi_\alpha(l - \frac{k_1 - k}{2})(E - E_\alpha)(E - E_{\alpha_1})$

$$\times g_0(l - \frac{k_1 - k}{2}, E)V(l - \frac{k_1 - k}{2}, l + \frac{k_1 - k}{2}, E)$$

$$\times g_0(l + \frac{k_1 - l}{2}, E)\psi_{\alpha_1}^*(l + \frac{k_1 - l}{2})\psi_{\alpha_1}(k_1)$$

again in the limit approximately

and $k, k_1 \to 0$ find

$$V_{eff} \cong \frac{V_0}{1 - \Omega\rho\left(\frac{E - E_\alpha}{E}\right)^2}$$

As can be seen the renormalized is more attractive; this is consistant with our previous result. It is interesting to note that the effect is energy dependent; higher excited states are more affected then low lying ones. This is expected since the range of bound state wave functions increases with energy.

As a consequence the energy of excited states will increase and the absorbtion spectrum will shift into the blue.

Such effects have been observed by Albrecht and Scott in their study of Rydberg states of benzene solutions in hexane[12].

It should be noted that the previous calculation over estimates the blue shift since $\rho(R)$ is not constant and vanishes for $R \to 0$. The main qualitative result should however still obtain for better models.

The localizing effect of the pseudo potential is modified by electron-electron interactions. We will discuss this in the following.

The effect of electron-electron interactions on impurity states in crystaline, dielectric media has been studied by Kohn[13]. The result of his analysis is that the impurity levels are determined by the effective Schrodinger equation.

(33)
$$\left(\frac{p^2}{2m^*} + \frac{e^2}{\epsilon r} \right) \psi = E\psi$$

where m^* is the effective mass, ϵ the zero frequency limit of the dynamic electronic dielectric constant. We will study the problem for the case of a disordered molecular medium assuming that intermolecular electronic exchange can be neglected.

We again consider the one particle Green Function The diagramtic expansion for g has the form

here ● represents the impurity potential, in the Coulomb interaction. It is convenient to introduce a scattering amplitude $T_{kk}{}'$, whose diagramatic expansion is given by

To calculate the diagrams appearing in either expansion we assume that

(34)
$$g_0(x, x_1, E) = \sum_{R_n, \alpha} \frac{\psi_\alpha^*(x - R_n)\psi_\alpha(x_1 - R_n)\big(1 - \Theta(\epsilon_F)\big)}{E - E_\alpha + i\delta}$$
$$+ \sum_{R_n, \alpha} \frac{\psi_\alpha^*(x - R_n)\psi_\alpha(x_1 - R_n)\Theta(\epsilon_F)}{E - E_\alpha - i\delta}$$

each $\psi_\alpha(r - R_n)$ is assumed to be localized on the solvent molecule at the site R_n. Also since intermolecular exchange is neglected all diagrams contain only loops corresponding to a particular n.

We are interested in the assymptotic behavior of the effective potential in the medium V_{eff} as $r \to \infty$ and thus in the small momentum behavior of T_{kk_1}.

For the case of complete disorder we find that diagrams of the type

$$\bullet -\!\!\otimes\!\!- \bullet \;\; = \;\; -\!-\!(R_n)\!-\!- \;\; + \;\; -\!-(R_n)\!-\!(R_m)\!-\!- \;\; + \;\; -\!-(R_n)\!\doteq\!(R_m)\!-\!-$$

give
$$T_{kk_1} \sim \frac{\text{Constant}}{|k - k_1|^2}$$

while diagrams such as

$$\bullet -\,(R_n)\,\Big]^{--}_{--}$$

give
$$T_{kk_1} \sim \frac{\text{Constant}}{|k - k_1|}$$

this implies that

$$V(r) \sim \frac{1}{\epsilon r} + \frac{\beta}{r^2} + \cdots$$

The first class of diagrams corresponding to successive scattering from different solvent molecules is important in the high density limit; the second class of diagrams corresponds to repeated scattering from the same solvent molecule and thus contains multipole contributions.

As can be seen Kohns result will hold for disordered media only if the impurity wave function extends over many solvent molecules. If the impurity levels are strongly localized, multiple corrections to the effective potential will be important.

The above discussion can be summarized as follows.

If bridge effect can be neglected a medium composed of polarizable molecules will tend to delocalize impurity states, shift the impurity absorbtion spectrum into the red. On the other hand a solvent composed of weakly polarizable molecules will localize states and shift the spectrum into the blue. As a result the electronic tunneling factor should be enchanced in polarizable solvents.

In such solvents however, polarization energies can become important and the activation energy of the process must be considered.

We will discuss this below using a model similar to that originally introduced by Holstein in his study of Polaron Motion[14].

We consider a donor-acceptor system containing one electron interacting with a random array of oscillators.

The Hamiltonian of the system is

$$H = \frac{p^2}{2m} + V_A + V_D + \sum_n \frac{e^2(\vec{r} - \vec{R}_n) \cdot \vec{\xi}_n}{|r - R_n|^3} \qquad \text{a)}$$

(35)

$$+ \sum \frac{P_n^2}{2M} + \frac{M\Omega^2}{2} \xi_n^2 + \sum_{n,m} \frac{e^2 \vec{\xi}_n \cdot \vec{\xi}_m}{|R_n - R_m|^3} \qquad \text{b)}$$

$$+ \sum \left(\frac{e^2 \vec{\xi}_n \cdot (\vec{R}_n - \vec{R}_A)}{|R_n - R_A|^3} + \frac{e^2 \vec{\xi}_n \cdot (\vec{R}_n - \vec{R}_D)}{|R_n - R_D|^3} \right) \qquad \text{c)}$$

here a) is the Hamiltonian of the electron donor system and the interaction term. b) the Hamiltonian of the oscillator system and c) the interaction Hamiltonian between the oscillator system and the donor and acceptor ions.

Following Holstein we assume that the wave function of the system has the form

$$\psi = a(\xi_1 \ldots \xi_n) \psi_a(x) + d(\xi_1 \ldots \xi_n) \psi_d(x)$$

where ψ_d and ψ_a are suitable electronic wave functions localized on the donor and acceptor respectively.

Noting that the quadratic Hamiltonian describing the oscillator system can be diagonalized we find that a and b satisfy

(36)

$$\begin{pmatrix} E_a + \sum \left(\frac{P_n^2}{2M} + \frac{M\tilde{\Omega}_n^2}{2} \eta_n^2 + e\eta_n \mathcal{E}_{an} \right) & \Gamma \\ \Gamma & E_d + \sum \left(\frac{P_n^2}{2M} + \frac{M\tilde{\Omega}_n^2}{2} \eta_n^2 + e\eta_n \mathcal{E}_{dn} \right) \end{pmatrix} \begin{pmatrix} a \\ d \end{pmatrix} = E \begin{pmatrix} a \\ d \end{pmatrix}$$

Here $\tilde{\Omega}$ are the frequencies of the collective modes of the oscillator system. $\mathcal{E}_{an}$ and $\mathcal{E}_{dn}$ are the electric fields acting at the Nth site when the electron is on the acceptor and on the donor respectively. All these quantities are random variables since they depend on $\{R_n\}$.

Denoting the matrix Hamiltonian by H we can define a Green Function $\mathcal{G}$ by the equation.

(37)

$$\left(\frac{\hbar}{i} \frac{\partial}{\partial t} - H \right) \mathcal{G} = I \delta^4 (t - t_1, x - x_1)$$

The calculation of the transition rates can be done as before. We must now take into account that we are dealing with a large number of vibrational states characterized by some thermal distribution function. Assuming Boltzman statistics, first order perturbation theory gives

128

$$(38) \qquad \text{rate} \sim \frac{\Gamma^2}{\hbar} \exp\left\{ \frac{-(E_d - E_a - E_{Sol}(R_n))^2}{4E_{Sol}(R_n)kT} \right\} \text{ X } \frac{1}{(E_{Sol} \text{ KT })^{\frac{1}{2}}}$$

where

$$E_{Sol}(R_n) = \sum \frac{M\Omega_n^2}{2} \xi_n^2$$

the result is similar to MLDH. In our case we still have to average over R_n. As a consequence the temperature dependence of the rate will change if the distribution of the E_{Sol} is broad enough electron transfer at low temperatures can occur.

we find

$$(39)$$

$$\begin{aligned} \frac{\text{rate}}{T \to 0} &\sim \frac{\bar{\Gamma}^2}{\hbar} \int dE_{Sol} \exp\left\{ \frac{-(E_d - E_a + E_{Sol}(R_n))^2}{4E_{Sol}(R_n)kT} \right\} \rho(E_{Sol})\delta(E_a - E_d - E_{Sol}) \\ &\cong \frac{\bar{\Gamma}^2}{\hbar} \rho(E_a - E_d) \end{aligned}$$

were ρ is the energy density distribution function of the random solvent.

Higher order corrections will renormalize Γ and the coupling will become frequency dependent.

The qualitative featurers connected with this can be understood if we recall that the screened potential of an ion in the medium has the form

$$V_{eff} = \frac{V_0}{\epsilon(\omega)}$$

within the molecular orbital method we are using

$$(40) \qquad \Gamma_{eff} = \frac{E_a + E_d}{2} \int \psi_a^* \psi_d \, dVol + \frac{1}{\epsilon(\omega)} \int \psi_a^* \psi_d V_d \, dVol + \int \psi_a^* \psi_d V_a \, dVol$$

using this expression for Γ, first order perturbation theory for the rate gives addition terms of the form

$$(41) \qquad \sum_\gamma \exp\left\{ \frac{-(E_d - E_a - E_{Sol}(R_n) + \hbar\Omega_\gamma)^2}{4E_{Sol}(R_n)kT} \right\}$$

where Ω_γ are the frequencies of the collective modes of the medium.

As can be seen collective modes can assist the transfer process by supplying energy to overcome the polarization energy associated with the reaction.

DISCUSSION

We have developed a Green Function formulation of electron transfer reactions which is essentially a time dependent generalization of standard molecular orbital theory.

Because of this, it can be used to study the effect of the symmetry of orbitals, through bond transfer etc. The interactions with the solvent can be incorporated into the formalism once the initial and final donor and acceptor states in the medium are known. Regarding the role of the solvent we distinguish two cases.

In the limit where polarization effects are small the solvent molecules can act as a bridge if they are suitably oriented in space (with respect to donor and acceptor molecules) and have the appropiate electronic states. In such cases the rate of the reaction will depend upon the formation of a bridging structure. If no adequate states connecting donor and acceptor molecules exist, the solvent will act as a barrier (of magnitude $\rho \Omega \Delta E$) through which the electrons have to tunnel. Fluctuations in the barrier can then become important in determining the transfer rate. Implicit in the preceding is the assumption of well defined initial and final donor and acceptor states. The determination of these states is the main problem in using the formalism.

If the solvent molecules are polarizable two effects will be of importance. In the first place polar solvents will in general tend to delocalize the electrons and may therefore assist electron transfer. On the other hand the changes in polarization energy accompanying the reaction will lead to temperature dependent factors in the expression for the rate.

Both the range of wave functions and the polarization energies are affected by disorder in homogeniety of the solvent leads to non-homgeneous broadening of electronic levels. Such broadening will in general aid electron transfer and will after the temperature dependence of the MDLH expression.

In particular low temperature electron transfer can be made possible. Disorder in the solvent will also affect the formation of bridging structures resulting in a "percolation" mechanism of electron transfer.

Finally if solvent collective modes exist they can again affect the temperature dependence of the rate by supplying energy to the reaction.

To study this further requires an investigation of disordered polarizable systems, in particular the model of a Bose Spin Glass with short range spatial correlations.

ACKNOWLEDGMENT

I would like to thank John Lindsay, D. M. Mauzerall and T. M. Rice for useful discussions.

REFERENCES

1) E. Butler, S.L. Guberman and A. Dalgarno Phys. Rev. A 16 500 1977.

2) R. Johnson and M.A. Bondi Phys. Rev. A 18 996 1978.

3) W.R. Green, F. Wright, J. Young and S.E. Harris Phys. Rev. Letters 43 10 1979.

4) J.H. Black and A. Dalgarno Astrophysics Letters 15 79 1975.

5) L. Landau and I. Lifshitz Quantum Mechanics 2nd Ed. Pergamon Press

6) E.E. Nikitin and B.M. Smirnow, Sov. Phys. Usp 21 (2) 95 1978.

7) See for example J. Ullstrup Theory of Electron Transfer Reactions Springer Venlag 1979.

8) J. Calef and D. Wolyness Journal of Chem. Phys. 87 3387 1983

9) R. Hoffmann A. Inamura and W. Hehne Journ. Chem. Soc. 90 1499 1968.

10) E. Stein and H. Taube Journal Am. Chem. Soc. 1981 103 693 see also A. Beretan and J.J. Hopfield Journ. Am. Chem. Soc. 106 1584 1984.

11) See for example R. Kubo and M. Nagamiya Solid State Physics McGraw Hill Co. 1969.

12) E. Albrecht Symposium on Photo Physics and Photo Chemistry above 6eV Bombammes France 1985. Also T. Scott private communication.

13) W. Kohn Phys. Rev. 105 (2) 509 1957.

14) T. Holstein Annals of Physics 8 325 1959.

TIME-RESOLVED DIELECTRIC LOSS AS A PROBE OF INTERFACIAL ELECTRON TRANSFER DYNAMICS

M. R. V. Sahyun

Science Research Laboratory, 3M
3M Center, 201-2E-04
St. Paul, MN 55144 USA

INTRODUCTION

A variety of photochemical reactions can be catalyzed by dispersed semiconductor powders. The basis for so-called photocatalysis, in which absorption of light by the semiconductor promotes an otherwise thermo-dynamically favored process, is the generation of electrons and positive holes in the semiconductor. These carriers can then react with the dis-persion solvent or a reagent dissolved therein (1-3). The dynamics of the interfacial electron transfers to and from the solution-phase reagents are thus critical to the efficiency of light utilization in the process.

In a photocatalytic system, each particle along with its surrounding sphere of solvent and solute comprises a complete electrochemical cell, in which, under illumination, anodic and cathodic half-reactions occur. From the band bending which occurs in the semiconductor at the solution inter-face (2,4), we infer that anodic processes are facilitated with n-type semiconductors, while cathodic process naturally tend to occur on p-type materials. Some form of surface modification of the semiconductor is accordingly required in order to obtain the complete cell chemistry.

In the absence, particularly, of the less-preferred half-reaction, the other half-reaction may provide a pathway for surface recombination on the semiconductor, e.g. on an n-type material the product of the anodic half-reaction can capture a photoelectron from the conduction band to regenerate the starting molecule, with net consumption of both a hole and an electron. While not desirable for doing useful chemistry, this process allows one half-reaction to be isolated for study of its dynamics on the semiconductor surface.

We illustrate the strategy by reference to two studies on bulk recombination in doped semiconductors. In CdS powders doped with Cu(I) or Ag(I), the rate of decay of photoconductivity, i.e. the rate of disappear-ance of the majority carriers from the conduction band, is limited by the rate of trapping of positive holes at the dopant center (5,6). This result is consistent with a steady-state analysis of the recombination scheme proposed by Dussel and Boer (7) and subsequently elaborated by Hadley (8). On the other hand, in single crystal ZnO (9), positive holes rapidly equilibrate between the trapping state(s) and the valence band, so that the traps buffer the positive hole concentration; binary recombination then follows pseudo-first order kinetics:

$$k_{obs} = k_r[T]exp(-\Delta G_T/k_0 T) \qquad [1]$$

where k_r is the second order recombination rate constant, [T] is the volume concentration of recombination centers, and ΔG_T is the free energy change associated with hole trapping (not to be confused with the "vertical" or optical trap depth, E_T, which is an enthalpy change (9a)).

Should a hole scavenging reagent adsorbed to an n-type powder behave in the manner of the Ag(I) or Cu(I) centers in CdS, measurement of the contribution of surface recombination to the overall photoconductivity decay rate can yield the frequency of interfacial electron injection from the reagent into the valence band of the semiconductor. On the other hand, if the physics observed in the ZnO single crystals is the more appropriate model, then perturbation of the recombination kinetics by the redox reagent will yield thermodynamic information about the surface state formed by the adsorbed reagent. Both pieces of information are of potential significance to the understanding and engineering of photocatalytic systems. It is tempting to associate the former limiting case with relatively inefficient non-adiabatic electron transfer at the interface, involving, presumably, tunneling (10), while the latter case should correspond to adiabatic interfacial electron transfer.

The time resolved dielectric loss technique (TRDL), which is a flash photolysis-microwave photoconductivity experiment, provides a probe which allows us to "count" conduction band electrons as a function of time after their photogeneration. It has been successfully employed in the study of the dynamics of dry II-VI powders (5,6,11), and has been adapted to the study of semiconductor electrode function (12). Recently, Warman and co-workers (13) demonstrated that TRDL is applicable to the photophysics of solvent dispersed powders, and we extended these results to aqueous dispersions of CdS (14).

Conceptually the experiment is related to, and complementary to, flash photolysis-esr (15,16); TRDL probes the free, mobile majority carriers in the conduction band, while the latter probes trapped carriers in well-defined sites. Operationally, the two techniques differ in the absence and presence of a magnetic field, respectively. The schematic of the TRDL experiment is shown in Fig. 1.

EXPERIMENTAL

Samples of the semiconductor powders are ultrasonically dispersed in the appropriate solvent or solution. We introduce the dispersions (ca 0.3 ml) into 3 mm thin-wall quartz tubes, which are then placed in the X-band microwave cavity of the apparatus of Fig. 1. Where required, we purge the dispersion of oxygen by bubbling dry nitrogen. The samples are exposed with the unfiltered output of a Xe flash lamp, of pulse duration ca 3 μs. The Schottky diode detector is in a bridge circuit, so only displacements of microwave power density in the cavity, i.e. conduction band carrier concentration (17), from the dark, equilibrium value are observed. We follow decays of the free carrier population for at least two half-lives.

The sub-micron French process ZnO and micron-sized raw CdS powders came from New Jersey Zinc Co. and General Electric Co., respectively. Doped CdS powder samples were prepared in these Laboratories. Organic solvents and reagents were products of commerce, purified as required. We used standard analytical techniques, such as melting point, nmr and tlc, to confirm the identity and purity of the materials.

RESULTS AND DISCUSSION

System I. Zinc Oxide-Carboxylic Acids (18)

Zinc oxide provided one of the first examples of photocatalytic water cleavage (19). Single crystal ZnO has been used as a photoconductive electrode to effect the Kolbe reaction (20), and Kraeutler and Bard (21) have reported a photocatalytic Kolbe reaction with TiO_2 as the semiconductor. Carboxylic acids and anhydrides have been used to modify the photophysical properties of ZnO for electrophotographic applications (22).

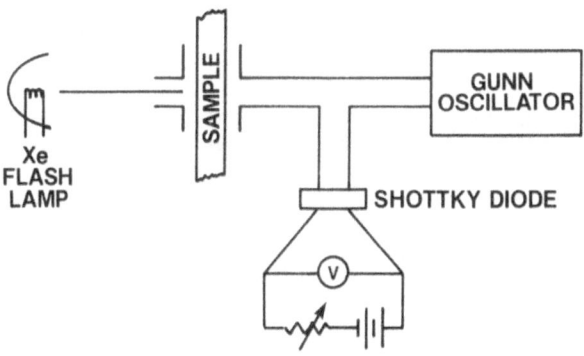

Figure 1.-Schematic of the TRDL experiment.

We found no evidence in either TRDL or preparative scale experiments of a photo-Kolbe process occurring in our systems. However, perturbation of the photophysics of the ZnO was observed by TRDL on xylene dispersions containing 0.001-0.25 M acetic, stearic, phenylacetic, cinnamic or benzoic acid. We associate these observations with the chemisorption of the carboxylic acid on the (alkaline) ZnO surface, presumably, by Zn carboxylate formation. In deoxygenated xylene dispersion, the photoconductivity decay in flash exposed ZnO is second order, typical of binary recombination. On admission of oxygen, the decay becomes pseudo-first order, with lifetime, Υ, ca 1.4 ms. This phenomenon is attributed to surface recombination mediated by chemisorbed superoxide ion (23). Faster pseudo-first order decays, typically 0.3 ms, are observed in the presence of the acids; when both acid and oxygen are present, the first order decay plots are bifurcated, as shown in Fig. 2.

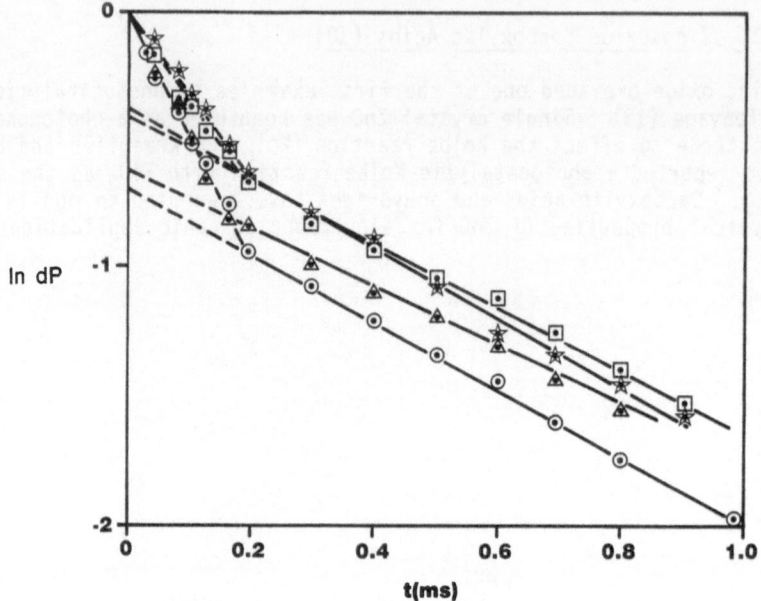

Figure 2.-First order photoconductivity decay plots for flash exposed ZnO in xylene dispersions containing various levels of stearic acid.

The lifetime, τ_1, of the faster decay component, obtained only in the presence of the acids, proved independent of (a) acid concentration, and (b) to a first approximation, chemical structure of the acid. Both the slope ($1/\tau_2$) and intercept of the least squares line correlating points associated with oxygen mediated decay, varied with the structure and concentration of the acid. From this variation in the lifetime of the oxygen meidated component of surface recombination, we infer that oxygen and the acids chemisorb competetively.

Consider the photophysical mechanism implied by eq. [1]. Given that the free energy of trapping, $\Delta G_T = E_T - T \Delta S_T$, and that ΔS_T scales with the occupancy of the sites available for chemisorption of the surface recombination catalyst (9a),

$$k_{obs} = k_r[S_0]\exp(-E_T/k_0 T) \qquad [2]$$

so that the observed pseudo-first order recombination rate constant, k_{obs} = $1/\tau_1$, should be independent of the solution concentration of the acid, as observed. The enthalpy of activation of the acid mediated surface recombination, ΔH^*, then corresponds to E_T, and, measured on representative dispersions incorporating the various acids, is 0.24 $\pm$ .05 eV, while E_T, estimated from eq. [2], using $k_r = 4 \times 10^{-9}$ $cm^3 s^{-1}$ (9) and $[S_0]$ = 10^{17} cm^{-3}, (18), is 0.27 $\pm$ .02 eV for all the acids studied. These observations confirm the hypothesis that photoholes equilibrate rapidly between the Zn carboxylate surface states and the valence band. Given the valence band (bulk) ZnO at ca +2.6 V <u>vs</u> SCE, this trap depth defines the redox potential of the chemisorbed acids as 0.5 - 0.7 V less reducing than

typical carboxylic acids in solution (23a), consistent with correspond-
ingly exergonic compound formation on adsorption to the oxide surface.

Kinetically, this equilibration implies that photohole capture by
the chemisorbed carboxylate species must occur with a frequency ca 8 x
10^{11} s^{-1}, representative of the adiabatic electron transfer regime. The
attempt to escape frequency, which determines the rate of reemission of
the surface trapped holes into the valence band, must accordingly be ca 2
x 10^{6} s^{-1}, somewhat slower than in, e.g. CdS (24). The reason for our
failure to observe any evidence of Kolbe reaction products from the photo-
lysis of ZnO with sorbed carboxylic acids, even phenylacetic, is now
readily apparent: the rate of decomposition of the carboxyl radical
cannot compete with the rate of recapture of an electron from the ZnO,
i.e. hole reemission into the valence band.

System II-Aqueous CdS (14,25)

Photocatalysts in aqueous environment are of interest, as the
possible exploitation of photocatalysis in solar energy conversion by
water splitting has provided considerable impetus for the recent activity
in this field (4,19,26,27). In the TRDL experiment, however, water
absorbs all the microwave energy in the cavity, thus spoiling the probe
system. We found, however, that we could measure TRDL on flowing thin
films of aqueous dispersions of, particularly, CdS powders, formed when
the dispersion was allowed to flow down the inside bore of a 3 mm quartz
tube passing through the microwave cavity.

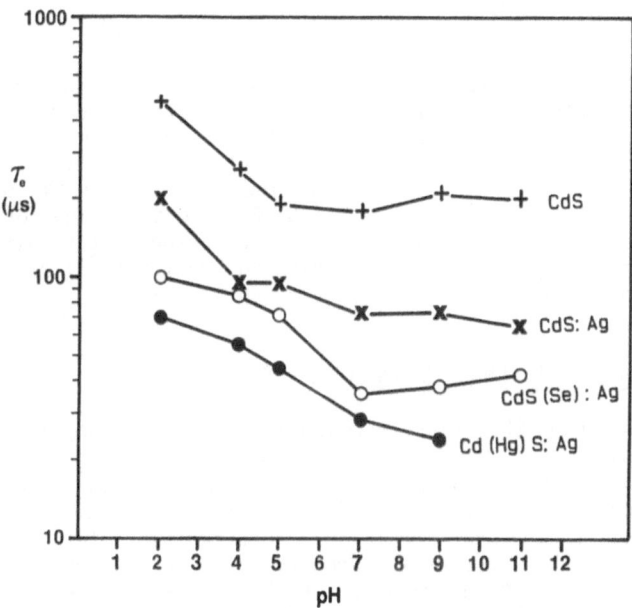

Figure 3.-Recombination activity profiles for various CdS preparations in
aqueous media: pH dependence of recombination lifetimes.

We studied five different CdS dispersions in aqueous dispersions buffered at a series of pH's from 2 to 11. The pH dependence of the TRDL free photoelectron lifetimes is shown in Fig. 3. From this pH dependence we infer that the surface photohole capture, in which adsorbed OH^- may be the active species (27), determines the surface recombination rate. Unlike the ZnO-carboxylic acid case, it is not limited by the supply of positive holes.

Surface recombination may be more complex in this case, as the experiments were done in air. Photoelectrons may be captured from the CdS conduction band by O_2 to form superoxide ion (15,16), so that recombination actually occurs in solution, between the superoxide ion and the hydroxyl radical formed on hole capture. By analogy to the photoelectrocheimstry of a TiO_2 membrane (27a), we expect that the cathodic (electron capture) process occurs on the more intensely illuminated face of an individual CdS particle, while the anodic (hole capture) process occurs on a less intensely illuminated face. In support of involvement of a cathodic half-reaction, surface recombination in moist acetonitrile was completely suppressed on deoxygenation of the dispersion. Neither iodide nor sulfide ion perturbed the recombination process in the aq. CdS dispersions, as would be expected if photocorrosion of CdS were involved (28).

In the dry, doped powders, bulk recombination at the Ag(I) and Cu(I) centers is also pseudo-first order (see above) with time constant τ_o, measured on the dry powder. The contribution of surface recombination in the dispersions, with pseudo-first order rate constant k_{pt}, is then

$$k_{pt} = (1/\tau_{obs} - 1/\tau_o) \qquad [3]$$

The actual interfacial hole trapping, i.e. electron injection, frequency, k, can be extracted on treatment of the data by Langmuir-Hinshelwood kinetics, modified to allow for adsorption of OH^- at sites of distributed activity, i.e. according to a Freundlich, rather than a Langmuir, isotherm. Thus

$$k_{pt} = kK[OH^-]^n/(1 + K[OH^-]^n) \qquad [4]$$

where K is the average equilibrium constant for adsorption, and n is the exponent describing the spread of the distribution of adsorption site reactivities. The results of this analysis are given in Table I, where it can be seen that only k, and not the adsorption parameters vary with the CdS preparation. We also take the magnitude of k, which is comparable to that observed by Moser and Graetzel (29) for electron transfer from the conduction band of TiO_2, as indicative of a non-adiabatic pathway. The variation in k then suggests a variation in the tunneling barrier geometry among the CdS preparations.

System III.-CdS with Phenylhydrazines (30)

We also found that surface recombination in xylene dispersed CdS could be mediated by added phenylhydrazines. Variation in the substituent(s) on the aromatic ring of this reagent then allows variation in the ionization potential, hence oxidation potential, of the electron donor, and we can see if a linear free energy relationship obtains for surface recombination. The slope of the Bronsted plot describing such a correlation could, in turn, be diagnostic for the adiabaticity of the interfacial electron transfer (31). We have not yet achieved this goal, owing to lack of reliable oxidation potential data on the phenylhydrazines.

Table I.-Kinetic parameters for surface recombination in aq. CdS.

Material	n	$K(M^{-n})$	$k(s^{-1})$	r^a
CdS	0.15	13.5	1.1×10^4	0.89
CdS:Ag	0.14	10	1.4×10^4	0.95
$Cd_{.99}Hg_{.01}S$	---	---	1.8×10^{4b}	---
$Cd_{.98}Hg_{.02}S$:Ag	0.16	9.5	4.0×10^4	0.93
$CdS_{.85}Se_{.15}$:Ag	0.14	8.0	5.0×10^4	0.98

a) Correlation coefficient for best fit of eq. [4] to
experimental data, using parameters given.
b) Estimated from two data points using average values of
0.15 and 10 for n and K, respectively.

Consequently, we undertook preliminary analysis of this system in terms of a Hammett $\rho\sigma$ treatment, where σ is a substituent constant, describing the electronic effects of the substituent in the aromatic ring, and ρ indicates the sensitivity of the rate determining step in a reaction to these effects (32). For the para-substituted phenylhydrazines listed in Table II, we used σ^+, which is known to scale with the ionization potentials of benzyl radicals (33), while we used σ_I for the meta-substituted compounds.

The CdS powder was not deliberately doped; in absence of a solute which mediated surface recombination, decay of photoconductivity, detected by TRDL, was exclusively second order. In the presence of the phenylhydrazines, and, to a first approximation, independent of their concentration, the decay became pseudo-first order. The lifetimes are also listed for three temperatures in Table II. The trend in τ's with σ is consistent with oxidation of the phenylhydrazine being the rate determining step in surface recombination.

This inference was confirmed in an experiment in which a $Zn_{.55}Cd_{.45}S$:Ag powder (34) was substituted for CdS. Surface analysis of this powder (ESCA) indicates that it has an oxide layer ca 20 Å thick, while the oxide layer on the CdS sample is only a few atomic layers thick. This thick oxide layer should block the electron transfer from the conduction band to the oxidized, adsorbed reagent, thus obviating surface recombination in the Zn(Cd)S. Indeed no significant contribution of surface recombination to the photoconductivity decay in Zn(Cd)S could be observed, either in aqueous environment or in xylene dispersions containing phenylhydrazine. On the other hand, carboxylic acids chemisorbed to the oxide surface of the Zn(Cd)S much as to the ZnO, and perturbed the photoconductivity decay in the Zn(Cd)S much in the same manner, as well, indicating that the blocking surface layer on the Zn(Cd)S did not interfere with the photophysics hypothesized to obtain in the ZnO-carboxylic acid system, i.e. the oxide layer blocks cathodic, but not anodic, processes.

Table II.-Photoconductivity decay time constants by TRDL for raw CdS in xylene solutions of substituted phenylhydrazines, $YC_6H_4NHNH_2$.

Y	σ	@ T =	0^0	23^0	42^0
p-methoxy	-0.78		0.80 ms	0.75	0.73
p-methyl	-0.31		1.02	0.89	0.77
H	0		1.18	0.98	0.88
p-chloro	+0.11		---	1.05	---
m-methoxy	+0.25		---	1.14	---
m-chloro	+0.47		---[a]	1.04[b]	0.84
m-fluoro	+0.52		1.48[b]	1.13[b]	0.80

a) Only second order decay observed.
b) These data do not agree with those reported in the preliminary account of this work, ref. (30). We digitally filtered the TRDL signal to improve signal-to-noise ratio, and to allow the decays to be monitored through their fourth half-life, thereby excluding interference from the binary recombination process.

On inspection of the data of Table II it occurred to us that the initial expectation of a linear free energy relationship might be naive: the substituent coefficient scales with the ionization potential, which is an enthalpy, not a free energy, and the entropy of surface oxidation of the various phenylhydrazines might not be a simple function of the substituent on the adsorbed reagent, insofar as reagents of different ionization, hence redox, potential may sample the semiconductor density of states function at different points. Accordingly, we turned our attention to enthalpies of activation, which, from the so-called "first order" Marcus equation can be shown (35) to relate to the enthalpy change on reaction, hence ionization potential of the adsorbed reagent, as

$$\Delta H^* = \lambda_h + \Delta H/2 \qquad [5]$$

where λ_h is an enthalpy of reorganization, so that

$$d\Delta H^*/d\sigma = -1.15 \, k_o T \qquad [6]$$

From eq. [6] and the data as plotted in Fig. 4 we obtain ρ = -2.3 ± 0.6, again consistent with the involvement of a one-electron oxidation of adsorbed phenylhydrazine reagent as the rate determining step in surface recombination. The relatively small (compared with, e.g. ρ = -19 for the gas phase ionization of benzyl radicals (33)) absolute value of reflects, we believe, considerable stabilization of the product phenylhydrazine radical cation by the polar environment of the (primarily CdO) microcrystal surface.

Dogonadze (36) derived that the non-adiabatic electron transfer frequency, k, at the high temperature limit is given by

$$k = (k_0T/h)\mathcal{K}\exp(-\Delta H^*/k_0T)\exp(\Delta S^*/k_0) \qquad [7]$$

where $\mathcal{K}$ is the transmission coefficient for the probability of crossing betwen reactant and product potential surfaces at their intersection. Based on the analysis of MItchell (9a), we estimate (30) the $T\Delta S^*$ term in the free energy of activation as -0.17 eV. Assuming $k = 1/\tau$, i.e. that the surface sites are saturated with the phenylhydrazine at the experimental concentrations, and $\Delta H^* = 0.05$ eV, we find $\mathcal{K}$ of the order of 10^{-6}, consistent with the assumption of non-adiabaticity.

Where electron tunneling obtains, $\mathcal{K}$ takes the form (37)

$$\mathcal{K} = \exp[-(r - r_0)/a] \qquad [8]$$

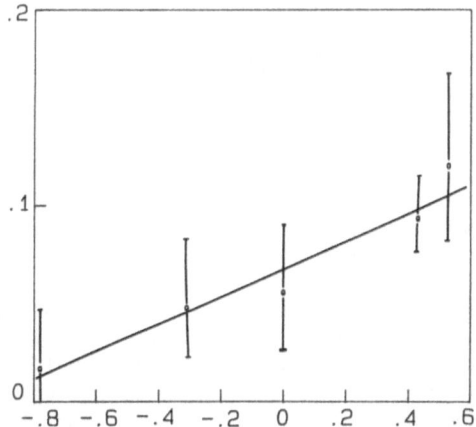

Figure 4.-Enthalpies of activation for surface recombination in CdS mediated by substituted phenylhydrazines, as a function of the phenyl-hydrazine substituent constant, σ.

where r is the barrier thickness, r_0 is the close-contact distance (at which electron transfer becomes adiabatic) and a describes the height and shape of the barrier. For typical values of $r_0 = 6$ Å and a = 0.8 Å, we estimate r to be ca 17 Å, somewhat thicker than analytical evidence suggests the oxide surface layer to be. The product kr is then the electro-chemical rate constant, of the order of 2×10^{-4} cm s^{-1}, and an order of magnitude smaller than observed by Moser and Graetzel (29) for the reduction of methyl viologen by conduction band electrons from TiO_2.

CONCLUSIONS

We have used the flash photolysis-TRDL technique to observe recombination in semiconductor powders mediated by adsorbed reagents. There are two limiting mechanistic cases. In the first, exemplified by carboxylic acids on ZnO, surface states formed by chemisorption of the reagent buffer the concentration of valence band photoholes, so that binary recombination becomes pseudo-first order. This photohole equilibrium requires interfacial electron injection from the adsorbed reagent to the valence band to occur on the time scale of, at most, tens of picoseconds, i.e. in the adiabatic regime. Compound formation, Zn carboxylate salts in this case, and the homogeneity of the semiconductor appear crucial to observation of this adiabatic interfacial process. The high frequency, adiabatic electron transfer pathway may not necessarily be desirable from the point of view of photocatalysis: the same factors which facilitate the electron injection process also facilitate the back transfer, which occurs here on the microsecond scale, faster than the oxidized surface species can undergo observable chemistry.

The second mechanistic regime involves adsorption of reagents, including OH$^-$ and substituted phenylhydrazines on CdS, which scavenge photoholes in the rate determining step in surface recombination. In the aqueous CdS case, the role of hydroxide ion can be described in terms of Langmuir-Hinshelwood kinetics, modified to account to adsorption at a distribution of sites. The interfacial electron transfer frequency in this case is of the order of 10^4 s^{-1}.

Comparable electron transfer frequencies are observed when the surface recombination is mediated by substituted phenylhydrazines. The enthalpies of activation of surface recombination depend on the phenylhydrazine substituent, amenable to analysis in terms of a Hammett treatment with ρ = -2.3, consistent with oxidation of the phenylhydrazine in the rate determining step. We estimate a transfer coefficient of ca 10^{-6}, consistent with non-adiabaticity, involving tunneling through a potential barrier ca 17 Å thick. This sort of chemistry cannot be observed with a Zn(Cd)S powder bearing a blocking oxide layer; the oxide layer blocks the cathodic, but not necessarily, the anodic half-reaction in surface recombination.

ACKNOWLEDGMENT

The author thanks Mr. Ronald G. Brisbois for assistance with much of the experimental work, done in partial fulfillment of the requirements for the B.S. degree in Chemistry at Hamline University, St. Paul, MN. Mr. G. H. Dierssen of these Laboratories was responsible for preparation of the doped CdS powders (System II). Finally, we acknowledge helpful and provocative discussion with Profs. David F. Ollis, North Carolina State University, and Nick Serpone, Concordia University, Montreal, Quebec.

REFERENCES

1   A. Henglein, Pure and Appl. Chem. 56, 1215 (1984).
2   J. Manassen, D. Cahen, G. Hodes, and A. Sofer, Nature 263, 97 (1976).
3   M. A. Fox, Acc. Chem. Res. 16, 314 (1983).
4   A. Heller, Science 223, 1141 (1984).
5   D. H. Klosterboer, W. L. Meissner and M. R. V. Sahyun, J. Appl. Phys. 54, 5161 (1983).
6   G. H. Dierssen and M. R. V. Sahyun, J. Appl. Phys. 56, 1647 (1984).

7   G. Düssel and K. W. Boer, Phys. Status Solidi 59, 275, 291 (1970).
8   H. C. Hadley, Jr., MS Thesis, Univ. Delaware, 1971.
9   M. Nitzan and Z. Burshtein, Phys. Status Solidi 58, K141 (1980).
9a   J. W. Mitchell, Photogr. Sci. Eng. 27, 96 (1983).
10   A. J. Nozik, D. S. Boudreaux, R. R. Chance, and F. Williams, Adv. Chem. 184, 155 (1980).
11   S. S. Collier, A. K. Weiss and R. F. Reithel, Photogr. Sci. Eng. 20, 54 (1976).
12   R. A. Bogomolni, H. Tributsch, G. Petermann, and M. P. Klein, J. Chem. Phys. 78, 2579 (1983).
13   J. M. Warman, M. P. deHaas, M. Graetzel, and P. P. Infelta, Nature 310, 306 (1984).
14   M. R. V. Sahyun, Chem. Phys. Lett. 112, 571 (1984).
15   J. R. Harbour and M. L. Hair, J. Phys. Chem. 81, 1791 (1977).
16   F. D. Saeva, G. R. Olin and J. R. Harbour, J. Chem. Soc., Chem. Commun., 401 (1980).
17   R. J. Deri and J. P. Spoonhower, Photogr. Sci. Eng. 25, 89 (1981).
18   M. R. V. Sahyun and R. G. Brisbois, Electrochimica Acta, in press.
19   A. A. Krasnovskii and G. P. Brin, Dokl. Akad. Nauk. SSSR 147, 656 (1962); 168, 1100 (1966).
20   S. R. Morrison and T. Freund, J. Chem. Phys. 47, 1543 (1967); P. A. Kohl and A. J. Bard, J. Amer. Chem. Soc. 99, 7531 (1977).
21   B. Kraeutler and A. J. Bard, J. Amer. Chem. Soc. 100, 2239 (1978).
22   I. A. Akimov, K. B. Demidov, L. N. Ionov, and T. I. Povkhan, Phys. Status Solidi 29, 359 (1975).
23   L. Grossweiner, Photochem. Photobiol. 8, 411 (1968); N. M. Beekmans, J. Chem. Soc., Faraday Trans. I 74, 31 (1978).
23a D. H. Geske, J. Electroanal. Chem. 1, 502 (1959/60); B. Kraeutler and A. J. Bard, J. Amer. Chem. Soc. 99, 7729 (1977).
24   A. Rose, RCA Review 12, 362 (1951).
25   M. R. V. Sahyun, Electrochimica Acta 30, 619 (1985).
26   K. Kalyanasundaram, E. Borgarello and M. Graetzel, Helv. Chim. Acta 64, 362 (1981).
27   K. Honda, Technocrat 16, 8 (1983).
27a Y. Yonezawa, R. Hanawa and H. Hada, J. Imaging Sci. (formerly Photogr. Sci. Eng.) 29, 171 (1985).
28   A. Fujishima, E. Sugiyama and K. Honda, Bull. Chem. Soc. Japan 44, 304 (1971); T. Inoue, T. Watanabe, A. Fujishima, K. Honda, and K. Kohayakawa, J. Electrochem. Soc., Electrochem. Sci. Tech. 124, 719 (1977).
29   J. Moser and M. Graetzel, J. Amer. Chem. Soc. 105, 6547 (1983).
30   M. R. V. Sahyun, Chem. Phys. Lett., 112, 571 (1985). See Note (b) to Table II for refinements to the experimental data realized since submission of this preliminary account.
31   S. Fukuzumi, C. L. Wong and J. K. Kochi, J. Amer. Chem. Soc. 102, 2928 (1980); 103, 7240 (1981).
32   L. P. Hammett, Physical Organic Chemistry, McGraw-Hill, New York (1940); E. M. Kosower, An Introduction to Physical Organic Chemistry, Wiley, New York (1968).
33   A. G. Harrison, P. Kebarle and F. P. Lossing, J. Amer. Chem. Soc. 83, 777 (1961).
34   D. K. Sharma, M. C. Palazzotto and M. R. V. Sahyun, J. Appl. Phys. 58, 936 (1985).
35   N. Agmon and R. D. Levine, Isr. J. Chem. 19, 330 (1980).
36   R. R. Dogonadze, in Reactions of Molecules at Electrodes, N. S. Hush, ed., Wiley-Interscience, London (1971).
37   J. R. Miller, J. V. Beitz and R. K. Huddleston, J. Amer. Chem. Soc. 106, 5057 (1984) and references cited therein; T. T. T. Li and M. J. Weaver, J. Amer. Chem. Soc. 106, 6107 (1984); R. A. Marcus, Int. J. Chem. Kinetics 13, 865 (1981).

CHARGE TRANSFER IN POLYPYRROLE AND ELECTROCHROMISM OF PRUSSIAN BLUE

Ephraim Buhks

Corporate Research Department, The BFGoodrich R&D Center

Brecksville, Ohio 44141

ABSTRACT

The data on electrical and optical properties of polypyrrole are discussed from the viewpoint of multiphonon charge transfer theory. This theory correlates the conductivity activation energy with the energy of light induced charge transfer, estimates the distance of elementary hop and the decay length of electronic tunneling matrix element, and explains the effects of polymer dopant concentration on electrical conductivity and near infrared spectrum.

Electrical and optical properties of Prussian Blue are explained by a similar mechanism, charge transfer hopping between mixed-valence iron centers. The dynamics of electrochromism in this material is shown to be associated with ionic diffusion and electron trapping process.

INTRODUCTION

Extensive experimental data have been collected in recent years on electrical and optical properties of conjugated doped conductive polymers.[1] Electrical conductivity in these materials changes by several orders of magnitude, depending on the dopant concentration, and reaches its maximum of 10^2-10^3 $\Omega^{-1}cm^{-1}$. Conductivity usually increases with temperature, although the functional dependence deviates from a simple Arrhenius law. A broad optical band in the near IR spectrum (~1 eV) appears under polymer doping. The intensity and the peak position of this band depend on the dopant concentration.

Conductive polymers exhibit electrochromic properties when electrochemically doped. The mechanism of electrochromism will be demonstrated using the example of Prussian Blue. This inorganic dye represents a model system characterized by charge transfer transitions between mixed-valence iron centers.[2]

The electrical properties of polypyrrole have been intensively studied.[3-5] In agreement with the published reports[3-5] the conductivity of oxidized polypyrrole, prepared by electrochemical deposition in our laboratory, was found[6] in the range of 10-70 ohm^{-1}cm^{-1} for the temperature range 200-400 K. The observed temperature dependence of the conductivity, σ, is characterized by a weak increase and saturation at 400 K. In Figure 1, this functional behavior was fit by the following equation

$$\sigma = AT^{-3/2} \exp (-E_a/kT), \qquad (1)$$

according to which the electrical conduction in polypyrrole proceeds via hopping processes of charge carriers between localized redox sites generated by anions introduced into the polymer by doping. The electron transfer rate temperatures dependence in a polymer is determined by the Franck-Condon factors of low-frequency vibrational modes. The observed conductivity activation energy, E_a = 60 meV, is a result of the interaction of a charge carrier with polarization of a polymer during an elementary jump. The activation energy is expected to be higher in polar polymers and it increases with the distance of the elementary jump, which is 6Å at the highest degree of polypyrrole oxidation (33%). The charge transfer theory[6] predicts that conductivity in polypyrrole increases exponentially with the doping concentration, p, reaching its maximum at the highest degree of oxidation, in agreement with the data,[7] according to

$$\sigma \sim p^{1/6} \exp(-2\alpha p^{-1/3}). \qquad (2)$$

This functional dependence of the conductivity in a polymer vs. degree of oxidation, displayed in Figure 2, is calculated for a typical value of $\alpha = 0.5$ Å^{-1} of the electronic tunneling decay parameter. A similar behavior was observed in O_2 dopped polypyrrole.[7]

The charge transfer mechanism was also verified by our study of the optical properties of polypyrrole.[6] The energy of the dielectric loss spectrum peak,[8] 0.25 eV, was found to be equal to the reorganization energy calculated from the value of the conductivity activation energy, $E_r = 4E_a$. The near infrared band of the absorption spectrum (Figure 3) and its position strongly depend on the degree of oxidation. These results agree with our theory which assigns this near infrared band to an optically induced charge transfer transition. Its intensity decreases exponentially with the elementary jump distance, $R = p^{-1/3}$, according to the functional form similar to Eq. (2). The peak energy is red shifted with decreasing doping concentration as $E \sim p^{1/3}$.

This theory also predicts[6] that the maximum conductivity of molecular doped polymers is in the range of 10^2-10^3 ohm^{-1}cm^{-1}. It explains the relative stability of heterocyclic conductive polymers, and provides a guideline for the synthesis of new conductive polymers.

PRUSSIAN BLUE

Prussian Blue, inorganic dye iron hexacyanoferrate, represents a well characterized model system for conductive polymers in our study of conductivity and optical transitions mechanisms associated with localized charge transfer processes.[9]

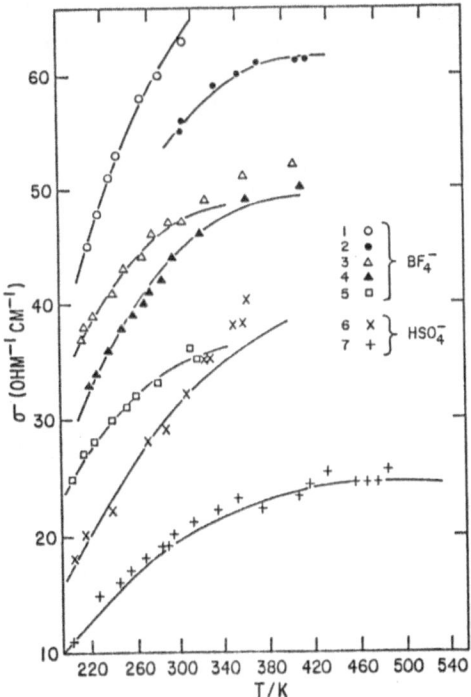

Figure 1. Temperature dependence of electrical conductivity of. poly-pyrrole doped with BF_4^- and HSO_4^-. The fitting curves were calculated according to Eq. 1 for E_a = 60 meV.

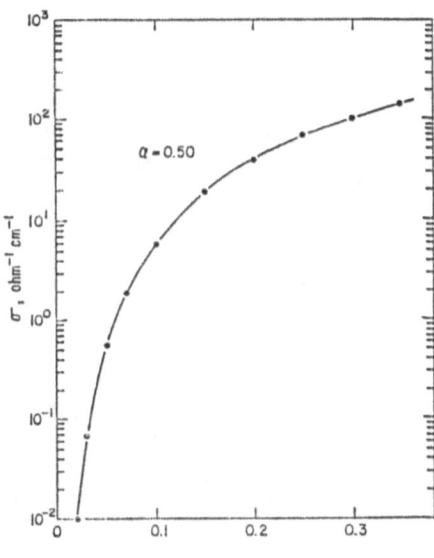

FRACTIONAL OXIDATION of POLYPYRROLE

Figure 2. Localized charge transfer model of electrical conductivity in conductive polymers as a function of polymer degree of oxidation.

Thin films (~1000Å) of iron hexacyanoferrate $Fe_4^{III}[Fe^{II}(CN)_6]_3$, Prussian Blue, are deposited on tin oxide/glass substrates electro-chemically.[10,11] The color of the film is bright blue and its absorp-tion spectrum displayed in Figure 4 is characterized by a broad band at 1.8 eV assigned to the intervalence transition[2]

$$Fe^{2+} + Fe^{3+} \xrightarrow{\nu} Fe^{3+} + Fe^{2+}. \qquad (3)$$

The peak energy equals approximately to the charge transfer reorganiza-tion energy, $E_r = 2.0$ eV, estimated from the distance of 5 Å between the iron centers.[2,12] The width at half height of the band, 0.7 eV, corresponds to a Gaussian line shape

$$\alpha(E) = \alpha_0 \exp[-(E-E_r)^2/4E_r kT]. \qquad (4)$$

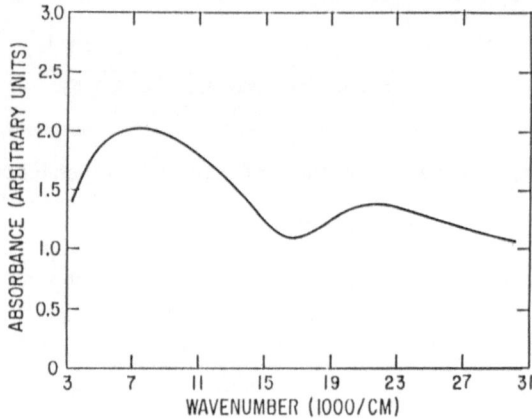

Figure 3. Absorption spectrum of the fully oxidized polypyrrole. The near infrared band corresponds to the light induced charge transfer between the localized sites.

The conductivity activation energy, $E_a = 0.5$ eV,[13] is close to $E_r/4$ which indicates multiphonon hole hopping process between the localized iron centers. Prussion Blue was found to be a p-type intrinsic semi-conductor characterized by a conductivity of 10^{-4} ohm^{-1}cm^{-1} at room temperature.[13]

Prussion Blue, like many other conducting polymers, exhibits electrochromic properties by bleaching its color when a reducing poten-tial is applied, as shown in Figure 4. Under electrochemical reduc-tion, the Prussion Blue film becomes transparent (Everett Salt), while the blue color is restored during oxidation. The bleaching process is explained in terms of electron trapping by the intervalence light absorbing center accompanied by K^+ attachment

$$Fe_4^{III}[Fe^{II}(CN)_6]_3 + 4e^- + 4K^+ \longrightarrow K_4Fe_4^{II}[Fe^{II}(CN)_6]_3. \qquad (5)$$

The decrease of the absorption intensity and the red shift of the peak energy with the reduction of Prussian Blue is quite similar to the changes in the near infrared absorption of polypyrrole and other conducting polymers during reduction. As in the case of polypyrrole, these effects can be explained by decreasing carrier concentration and increasing elementary jump distance of a photoinduced charge transfer between the localized centers.

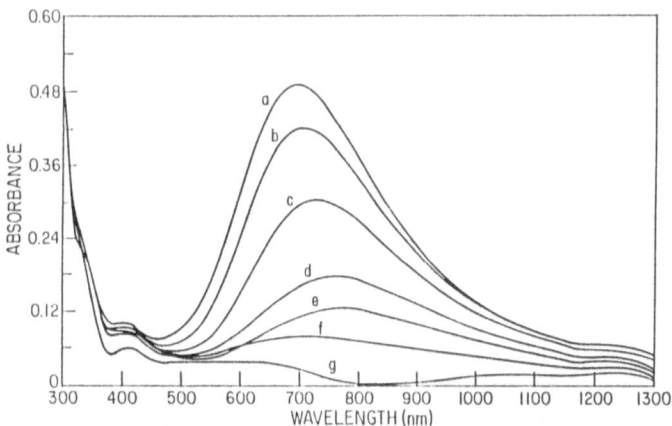

Figure 4. Absorption spectrum of Prussian Blue at various reducing potentials: (a) 0.6 V vs. SCE; (b) 0.3 V; (c) 0.25 V; (d) 0.20 V; (e) 0.15 V; (f) 0.10 V and (g) -0.10 V. The absorption band of Prussian Blue is assigned to intervalence charge transitions between Fe^{II} and Fe^{III} centers.

Dynamics of electrochromism of Prussian Blue was characterized using a He-Ne laser beam transmittance through the film and current measurement during reduction (-0.1 V vs SCE) and oxidation (0.6 V vs SCE) cycles in an electrolyte containing KCl. As shown in Figure 5, the kinetics of current and laser beam transmittance are quite similar. The typical switching time estimated from a 10% change in the transmittance is of the order of 100 ms.

The effect of the film thickness is demonstrated in Figure 6 where the transmittance kinetics data are plotted for 1000 Å and 2000 Å thick films. The increase in the switching time by a factor of 4 for the

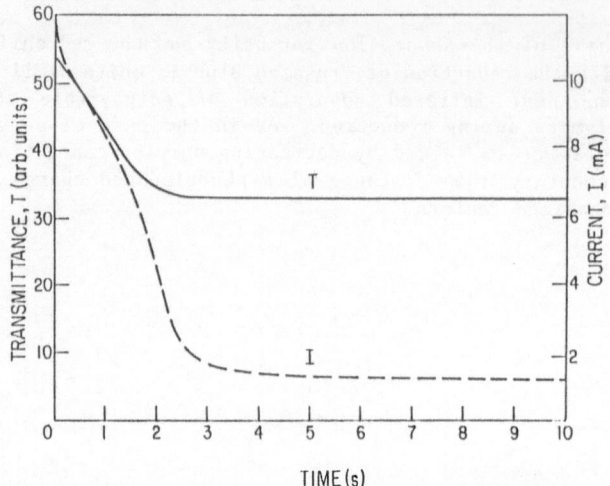

Figure 5.   Current (I) and He-Ne laser beam transmission (T) characteristics of Prussian Blue during the oxidation process.

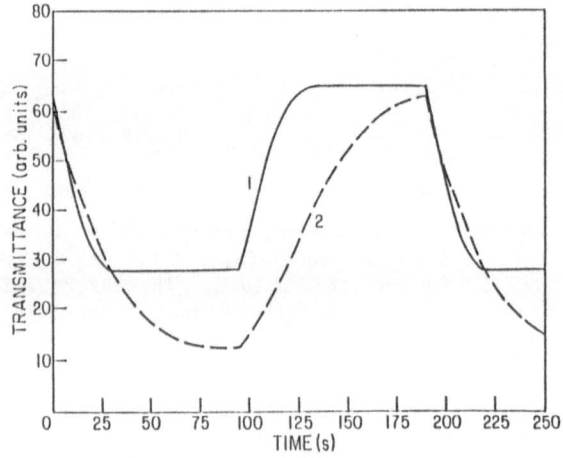

Figure 6.   Electrochromism dynamics of Prussian Blue.   Transmission
intensity of He-Ne laser beam through the Prussian Blue film during the
reduction and the oxidation cycles.   Increase of the switching time
with the film thickness is demonstrated by curves 1 and 2 which correspond to 1000 Å and 2000 Å thick films, respectively.

thicker film, indicates that ionic diffusion is the rate determining process.

The dynamics of increase in light transmission and decrease in current density during the bleaching (reduction) process was fit by a model assuming K^+ diffusion from the electrolyte into the Prussian Blue film.[9] Time dependence of the decrease in light transmission and decrease in current density during the coloration (oxidation) process was also fit by reverse diffusion of K^+ ions from the film into the electrolyte.[9] The contrast ratio time dependence is given by

$$T(t)/T(\infty) = (T_{PB}/T_{ES})^{\frac{8}{\pi^2} \sum_m \exp[-(2m+1)^2 t/\tau]/(2m+1)^2,}$$

(6)

where T_{PB} and T_{ES} are the transmittance of Prussian Blue and Everitt Salt, respectively; $\tau = 4\ell^2/\pi^2 D$, ℓ is the film thickness and D is the diffusion coefficients of K^+. Equation (6) describes the contrast ratio kinetics during the reduction process. A similar functional form is applied for coloration description during the oxidation process by interchanging T_{PB} with T_{ES}.

This model explains the dependence of the switching time of Prussian Blue electrochromic display on the film thickness and on K^+ diffusion coefficient which was found to be in the range of 10^{-8} - 10^{-9} cm²/s.

CONCLUSION

Multiphonon charge transfer theory explains the conductivity data and the optical spectra of conducting polymers, such as polypyrrole. The transport process in these materials is viewed as charge hopping between localized sites associated with dopant ions introduced by the oxidation process.

The estimated parameters of charge transfer, such as reorganization energy and electron tunneling decay length, predict fast electron transfer rates (~10 ps) in conducting polymers.

Inorganic model systems, such as iron hexacyanoferrate, Prussian Blue, exhibit electrical and optical properties similar to conducting polymers. The transport and optical absorption are explained in terms of intervalence charge transitions between Fe^{II} and Fe^{III} localized centers.

The electrochromic dynamics of Prussian Blue was studied by laser spectroscopy. The results indicate that the switching time is determined by ionic diffusion processes which depend on the film thickness and the ion diffusion coefficient.

ACKNOWLEDGEMENT

I would like to thank Drs. L. Traynor, I. Hodge and Mr. W. Gambogi for synthetic and characterization work.

REFERENCES

1. Recent reviews: (a) Proceedings of the Workshop on Synthetic Metals, Synth. Met., 9, 129-346 (1984); (b) Proceedings of the International Conference on Synthetic Metals, Mol. Cryst. Liq. Cryst., 117-121 (1985); (c) "Handbook on Conducting Polymers", Ed. T. J. Skotheim, Dekker, New York, 1985.

2. M. Robin, Inorg. Chem., 1, 337 (1962).

3. K. K. Kanazawa, A. F. Diaz, M. T. Krounbi and G. B. Street, Synth. Met., 4, 119 (1981).

4. G. B. Street, T. C. Clarke, M. Krounbi, K. Kanazawa, Y. Lee, P. Pfluger, J. C. Scott and G. Weiser, Mol. Cryst. Liq. Cryst., 83, 253 (1982).

5. A. Watanabe, M. Tanaka and J. Tanaka, Bull. Chem. Soc., Japan, 43, 2278 (1981).

6. E. Buhks and I. M. Hodge, J. Chem. Phys., 83, 5976 (1985).

7. J. C. Scott, P. Pfluger, M. T. Krounbi and G. B. Street, Phys. Rev. B, 28, 2140 (1983).

8. J. Yakushi, L. J. Lauchlan, T. C. Clarke and G. B. Street, J. Chem. Phys., 79, 4774 (1983).

9. L. Traynor, W. Gambogi and E. Buhks, J. Appl. Phys., to be published.

10. K. R. Rajan and V. D. Neff, J. Phys.Chem., 86, 4361 (1982).

11. K. Itaya, K. Shibayama, H. Akahosi and S. Toshina, J. Appl. Phys., 53, 804 (1982).

12. N. S. Hush, Progr. Inorg. Chem., 8, 391 (1967).

13. H. Inoue and S. Yanagisawa, J. Inorg. Nucl. Chem., 39, 1409 (1974).

ENERGY TRANSFER IN DOPED POLYMER FILMS

D.M. Hanson, Y. Lin, and M.C. Nelson

Department of Chemistry
State University of New York
Stony Brook, NY 11794-3400, USA

INTRODUCTION

Many studies of electronic excitation energy transfer in polymer solids have been conducted. Polymers have attracted attention because of their technological and fundamental importance and because they form a class of disordered materials. To our knowledge, no previous studies of direct donor - acceptor triplet state energy transfer in doped polymer solids have been reported. We recently began to study this phenomenon of sensitized phosphorescence for three reasons. First, data acquisition and analysis with a large laboratory computer system allow the many variables in a disordered three component system to be probed conveniently. These variables include the temperature, the concentration of the donor, the concentration of the acceptor, conditions of sample preparation, the wavelength of excitation, the wavelength of the phosphorescence, the intensity and time evolution of the donor phosphorescence, and the intensity and time evolution of the acceptor phosphorescence. Second, it appears that the concepts of Fractal Geometry /1/ allow properties of disordered systems to be characterized by a few parameters. To us, being able to characterize a disordered system by one or two numbers represents a significant advance, which should be exploited to improve our understanding of disordered materials. Finally, we are interested in the effects of electric fields on energy transfer processes, and large electric fields, more than one million volts per centimeter, can be obtained by applying a small potential drop across a thin polymer film.

A fractal is an irregular object that looks the same at any level of magnification or resolution. Such an object is said to be scale invariant or self-similar with dilation symmetry. Common examples of fractals are clouds, mountains, rivers, and coastlines. It has been recognized that disordered materials have properties of fractals, and concepts of Fractal Geometry have been applied to understand properties of multicomponent crystals, crystalline films, various aggregates, polymers, aerosols, colloids, surfaces, vycor and other glasses, filters, and zeolites. An important characteristic of fractals is that they can be described by three dimensionalities /1-3/: the Euclidean dimension, d, which is the dimension of the embedding space, the fractal dimension, d_f, which expresses a mass to size relation, and the spectral dimension, d_s, which expresses a connectivity relation. While d necessarily is an integer, d_f and d_s need not be integers. These three dimensions, however, are equal for an Euclidean object.

The concept of a fractal dimension is illustrated by considering how "dimension" can be defined by the relationship between distance (or length, L) and mass (or number of particles, N). If $N = cL$ where c is a constant of proportionality, we would say the system is one dimensional, if $N = cL^2$, the system is two dimensional, and if $N = cL^3$, the system is three dimensional. In general, if $N = cL^{d_f}$, then the dimension is d_f, which is called the fractal dimension and need not be an integer. The fractal dimension is relevant to our experiments on direct donor acceptor energy transfer because the rate of energy transfer depends upon the number and spacing of acceptor molecules around the donor.

One can imagine that several configurations are possible for molecules doped into a polymer matrix. Some possibilities are illustrated in Figs. 1-3. The molecules might form a random, isotropic distribution as appears to be the case for glassy matrices composed of small molecules, e.g. ethanol. The molecules might form microcrystals or other well defined aggregates, or the molecules might occupy sites in regions of space defined by the polymer structure. The objectives of the research described here are to characterize the phenomenon of triplet state donor acceptor energy transfer in doped polymer solids, to examine the applicability of the fractal dimension concept to this phenomenon, and to provide an answer to the question, What is the configuration of molecules doped into a polymer solid? The results reveal that the time evolution of the donor phosphorescence is not consistent with a random, isotropic distribution, that the guest concentration dependence is not consistent with the formation of microcrystals or aggregates, and that the data are consistent with using the concept of a fractal dimension to describe the distribution of the donor - acceptor system within the space defined by the polymer structure.

EXPERIMENTAL TECHNIQUES

Both the energy donor DBBP, 4,4'-dibromobenzophenone (ICN Pharmaceuticals), and the energy acceptor DBN, 1,4-dibromonaphthalene (Eastman), were purified by multiple recrystallization from ethanol. The polymer host, polystyrene (Aldrich), was used as received. The doped polymer films were prepared by mixing appropriate amounts of two solutions to obtain desired concentrations, expressed in weight per cent. For example, one solution consisted of 50 mg DBBP and 1 g of polystyrene dissolved in about 10 ml of toluene. A second solution consisted of 150 mg of DBN, 50 mg of DBBP, and 1 g of polystyrene dissolved in toluene. Films were made by placing a small amount of the final solution onto a 1 cm x 1 cm quartz plate that had been carefully cleaned. To control the rate at which the solvent evaporated, the plate was covered with a funnel and placed on top of a platform over a hot plate. The resulting films appeared visually to be uniform in thickness and optically clear. The films were about 0.1 mm thick. Variations in the thickness did not affect the energy transfer phenomenon.

A quartz plate with the doped polymer film was mounted on the cold finger of a cryorefrigerator and cooled to 16 K. As shown in Fig. 4, the energy level structure of this donor - acceptor system allowed the donor to be excited via the lowest singlet state without exciting the acceptor directly. This excitation at 386 nm was accomplished by using a nitrogen pumped dye laser with the BPQ dye. The phosphorescence was monitored through a double spectrometer using a photomultiplier tube. The data were collected by a PDP 11/23+ computer system through a CAMAC interface. Steady state measurements were made using a lock-in amplifier and an analog to digital converter. Transient measurements were made using a wide-band preamplifier and a 100 MHz transient digitizer with an averaging memory.

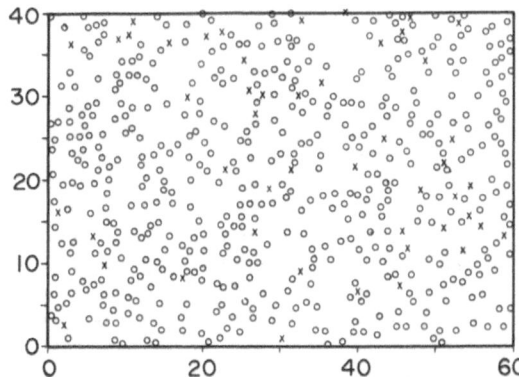

Fig. 1. Illustration of a random, isotropic distribution of donor molecules (*) and acceptor molecules (o). There is room in this two dimensional space for 2400 molecules. Here, 48 donors (2%) and 480 acceptors (20%) have been placed randomly without requiring that the molecules sit at lattice points.

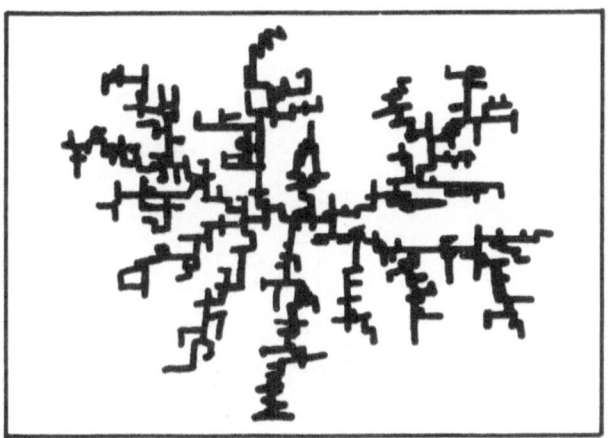

Fig. 2. Illustration of a diffusion limited aggregate in a two dimensional space.

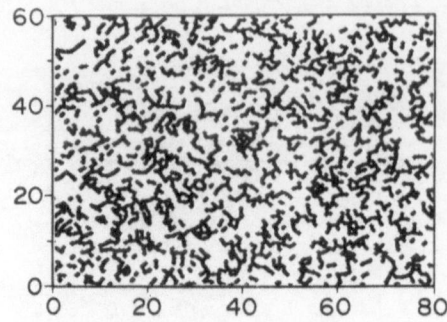

Fig. 3. The shaded areas in this illustration represent
regions of space defined by the polymer structure
(white area) that can be occupied by the donor -
acceptor system. The shaded areas were obtained by
randomly placing 960 molecules (40%) in this two
dimensional space as was done for Fig. 1, and then
neighboring molecules were connected together. This
procedure is similar to the construction of a
percolation cluster.

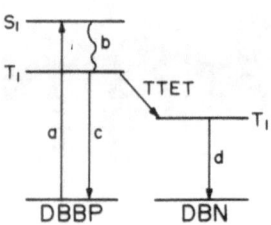

Fig. 4. Energy level diagram of 4,4'-dibromobenzophenone as
the energy donor and 1,4-dibromonaphthalene as the
energy acceptor. Laser excitation of DBBP via S_1 is
followed by rapid internal conversion to T_1 and then
energy transfer to T_1 of DBN or decay back to the
ground state.

THEORY

For the situation of direct (single step) donor - acceptor energy transfer, the decay of a donor molecules following pulsed excitation is described by the following equation,

$$(1) \qquad dP/dt = -k\ P - \Sigma_i\ W(r_i)\ P$$

where $P(t)$ is the probability that the donor is excited, and $1/k$ is the lifetime of the excited donor in the absence of acceptors. The last term describes the effect of energy transfer from the excited donor to surrounding acceptor molecules. The energy transfer rate is $W(r_i)$ where r_i is the position of acceptor i relative to the donor. For the case of triplet state energy transfer, the excitation transfer interaction generally involves electron exchange and drops off exponentially with distance /4/.

$$(2) \qquad W(r_i) = k\ \exp(gr_o - gr_i)$$

where g depends on the overlap of the donor phosphorescence and the acceptor absorption spectra, and r_o is the critical transfer distance /5/. The solution to Eq. 1 is

$$(3) \qquad P(t) = \exp(-kt - t\ \Sigma_i\ W(r_i))$$

for a particular configuration of acceptor molecules.

The measured quantity, however, is the ensemble average, $P_{av}(t)$, over all possible configurations /6,7/

$$(4) \qquad P_{av}(t) = \exp(-kt)\ \Pi_j(1 - p + p\ \exp(-t\ W(r_j)))$$

where p is the probability that an acceptor site is occupied by an acceptor. The product is over all sites. By taking the logarithm and changing the summation over sites to an integral over volume, Eq. (4) becomes

$$(5) \qquad \ln P_{av}(t) = -kt + 4\pi \int_o^\infty r^2\ D(r)\ \ln(1 - p + p\ \exp(-t\ W(r)))\,dr.$$

In a fractal structure, the density of acceptor sites, $D(r)$, is not a constant but depends on the fractal and Euclidean dimensions /7,8/.

$$(6) \qquad D(r) = D_o\ r^{d_f - d}$$

Using Eq. (6) in Eq. (5) gives

$$(7) \qquad \ln P_{av}(t) = -kt + 4\pi D_o \int_o^\infty r^{d_f-1}\ \ln(1- p + p\ \exp(-t\ W(r)))\,dr$$

To compare this theoretical result with the experimental data, we change variables $(y=gr)$, evaluate $\ln P_{av}(t)$ numerically, determine k from the phosphorescence decay when the acceptor concentration is zero, and determine the four parameters in the equation to fit the phosphorescence decay when the acceptor concentration is 10% by weight.

$$(8) \qquad \ln P_{av}(t) = -kt + C_1 \int_o^\infty y^{d_f-1}\ \ln(1 - ac + ac\ \exp(-kC_2 t\ \exp(-y)))\,dy$$

The four parameters are $C_1 = 4\pi D_o/g^{d_f}$, $C_2 = \exp(gr_o)$, a = the constant of proportionality between the site occupation probability and the known acceptor concentration in weight per cent, c, and d_f = the fractal dimension. Values for these parameters were obtained by examining 576 sets

157

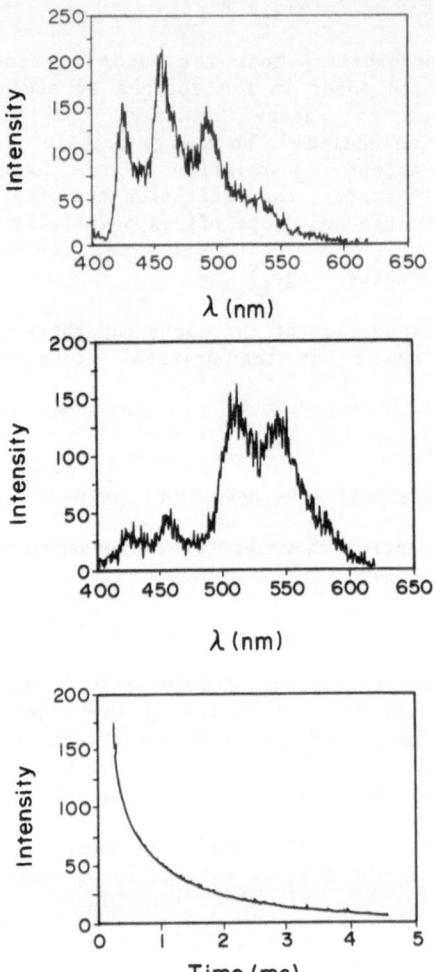

Fig. 5. Phosphorescence spectra of samples consisting
of 5% DBBP (top) and 5% DBBP with 10% DBN
(bottom). Decay of the DBBP phosphorescence
(454.1 nm) from 5% DBBP - 10% DBN sample.

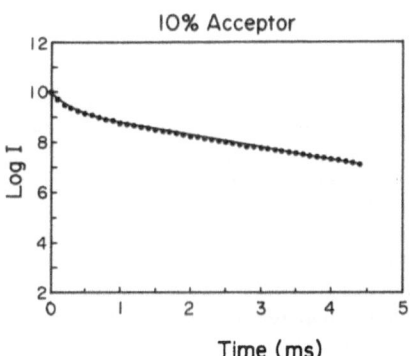

Fig. 6. Phosphorescence decay curves of DBBP at concentrations of 2% (*) and 5% (o) in the presence of 10% DBN. The fact that these curves are essentially identical indicates that multi-step donor-donor energy transfer to the acceptor is not significant.

of values within the following ranges for each of the parameters: $C_2 = 4.5$ to 8.0 in steps of 0.5, $k_f = 1.8$ to 3.0 in steps of 0.1, and $a = 0.004$ to 0.024 in steps of 0.004. For each case, C_1 was determined by a linear regression involving the experimental data and the integral in Eq. (8). The set of values giving the minimum for chi squared is $C_1 = 0.121$, $C_2 = 6.5$, $k_f = 2.5$, and $a = 0.012$

EXPERIMENTAL RESULTS AND DISCUSSION

In aromatic carbonyl molecules, intersystem crossing from the lowest singlet state to the lowest triplet state is very rapid, dominating other decay processes. The excitation pulse therefore populates the lowest triplet state of DBBP. Phosphorescence from this state is observed around 450 nm. Some DBBP molecules transfer the excitation energy to DBN, and sensitized phosphorescence from the lowest triplet state of DBN can be observed around 510 nm, see Fig. 5. These decay paths are shown in Fig. 4. No DBN phosphorescence could be observed in the absence of DBBP under our experimental conditions, even at the highest sensitivity settings of the apparatus.

The time evolution of the donor phosphorescence was measured as a function of both the donor and the acceptor concentration with good signal to noise as shown in Fig. 5. In the absence of acceptors, the phosphorescence decay appeared to be exponential with a lifetime of 2.75 ms (2%). As the acceptor concentration increased, the donor decay became fast and markedly nonexponential. Since our interest is in direct donor – acceptor energy transfer, it was necessary to find the concentrations for which donor – donor energy transfer (multi-step transfer to the acceptor) was not significant. Clearly one wants a low donor concentration and a high acceptor concentration, but there are practical limits. At very low donor concentrations, the signal to noise ratio is undesirably small. At high acceptor concentrations, visible aggregates form in the polymer film. In a series of steady state experiments, we found that the fraction of total phosphorescence that came from the acceptor appeared to be independent of the donor concentration provided the donor concentration was less than 5% and the acceptor concentration was 10% or higher. These results indicate the regime where donor – donor transfer is not significant.

Phosphorescence decay curves for DBBP at concentrations of 2% and 5% are plotted in Fig. 6 for a DBN concentration of 10%. The two sets of data are essentially identical. We therefore conclude that donor – donor transfer is not significant. Comparisons of the experimental data with theory were made using samples with DBBP concentrations of 2% and DBN concentrations of 10, 15, and 20%.

We first attempted to fit the decay curves to the conventional theory that adequately describes triplet state donor acceptor energy transfer in glassy matrices /5/. It was possible to obtain agreement between this theory and experiment at one concentration, but when the concentration was changed, the values for the parameters no longer were valid even though they should not be concentration dependent. It thus is clear that this theory is inadequate to describe triplet state donor acceptor energy transfer in these doped polymer solids.

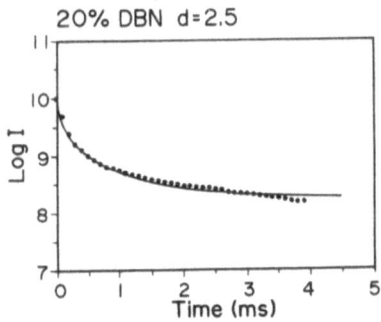

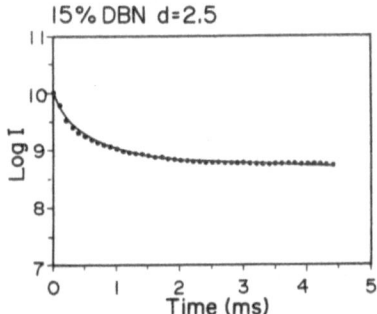

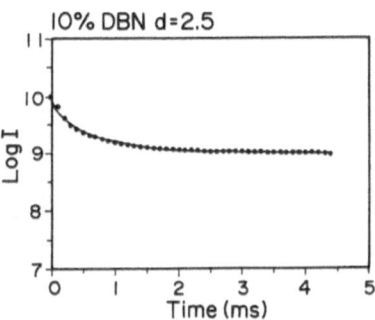

Fig. 7. Plots of ln P$_{av}$(t) + kt vs t for samples
containing 2% DBBP and 10, 15, and 20 % DBN.
The points represent the experimental data,
and the solid line shows the best fit to Eq.
(8) obtained using d$_f$ as a variable
parameter.

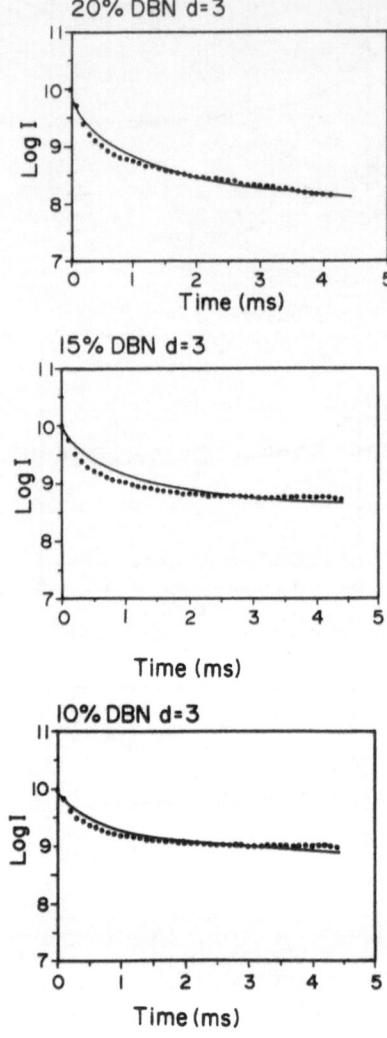

Fig. 8. Plots of $\ln P_{ay}(t) + kt$ vs t for samples
containing 2% DBBP and 10, 15, and 20 % DBN.
The points represent the experimental data,
and the solid line shows the best fit to Eq.
(8) obtained with d_f = 3.0.

It is our view that a significant omission of this theory lies in the assumption that the solid matrix is uniform and isotropic on a microscopic scale. While this assumption may be valid for glasses composed of small molecules, it does not seem reasonable for polymeric solids. The polymer occupies regions of space in a correlated fashion, and not all points in space are available to the dopant molecules. Thus, the number of acceptor molecules surrounding a donor should not scale as the distance cubed, characteristic of three dimensional space, but rather should scale as a power less than three, the fractal dimension. We therefore tried to fit Eq. (8) to the data. Eq. (8) is more accurate for higher acceptor concentrations and includes the concept of a fractal dimension. When this is done, it is found that a dimension of 2.5 fits the data better (chi squared is smaller) than a dimension of 3.0. Compare Figs. 7 and 8.

CONCLUSIONS

Triplet state excitation energy transfer from a donor molecule (4,4'-dibromobenzophenone) to an acceptor molecule (1,4-dibromonaphthalene) in a solid polymer matrix (polystyrene) at 16 K has been studied by monitoring the relative phosphoresence intensities and the time evolution of the donor phosphorescence as a function of the concentration of both the donor and the acceptor. The data reveal the concentrations for which multistep energy transfer is negligible. The best theoretical description of the experimental data is obtained for a fractal dimension of 2.5. This value for the fractal dimension means that the number of acceptor molecules surrounding a donor scales as the distance raised to the 2.5 power rather than scaling with the volume. The fractal dimension characterizes the space available for the donor – acceptor system within the polymer structure.

The significance of these results will be tested by comparison with corresponding data for this donor – acceptor system in a small molecule glass. For the small molecule glass, we expect about the same value for C_2, $d_f = 3.0$, and $a = 1$ when the concentration is expressed as the mole fraction. Furthermore, it may be possible to evaluate C_1 from the mass density and the spectral overlap of the donor and acceptor.

REFERENCES

1. B.B. Mandelbrot, "The Fractal Geometry of Nature," Freeman, San Francisco (1982).
2. S. Alexander and R. Orbach, J. Phys. Lett. 43, 625 (1982).
3. R. Rammal and G. Toulouse, J. Phys. Lett. 44, 13 (1983).
4. D.L. Dexter, J. Chem. Phys. 21, 836 (1953).
5. M. Inokuti and F. Hirayama, J. Chem. Phys. 43, 1978 (1965).
6. A. Blumen and J. Manz, J. Chem. Phys. 71, 4694 (1979).
7. A. Blumen, J. Chem. Phys. 72, 2632 (1980).
8. J. Klafter and A. Blumen, J. Chem. Phys. 80, 875 (1984).

VIBRATIONAL DYNAMICS OF HYDROGEN BONDED SOLIDS:

PHONONS AND OPTICAL DAMAGE

Thomas J. Kosic[*], Jeffrey R. Hill and Dana D. Dlott

School of Chemical Sciences
University of Illinois at Urbana-Champaign
505 South Mathews
Urbana, IL   61801

In the last few years we have used picosecond time-resolved coherent Raman scattering to study vibrational relaxation in low temperature molecular crystals. In 1983 we made the very interesting observation that the optical phonons (w = 30-100 cm^{-1}) of hydrogen bonded crystals had exceedingly long lifetimes, often in the nanosecond range [1]. We later studied a number of materials including amino acids, peptides, and acetanilide [2]. The long lifetimes imply extremely sharp Raman spectra with lorentzian linewidths in the range of .005 cm^{-1}.

Recently we have extended these investigations to the regime of extremely intense laser pulses with power densities of a few gigawatts/cm^2 [3]. In this regime, low temperature hydrogen bonded solids undergo the phenomenon of accumulated optical damage [4]. This process is illustrated in Figure 1. It is a two-stage process. In the first stage, the Incubation Period, defects are created and accumulated over thousands of laser pulses. In the second stage, the Destruction Period, the defect concentration becomes critical and optical damage appears as a cavity or inclusion in the crystal. The formation of damage is detected via light scattering, and typically takes place within one nanosecond [3].

Although we have not detected these defects spectroscopically, we have learned two essential facts about defect accumulation in low temperature acetanilide crystals: 1) defects are created by 4-photon ionization processes, and 2) the critical concentration of defects is very large, perhaps several percent, just prior to the formation of the damaged region [3].

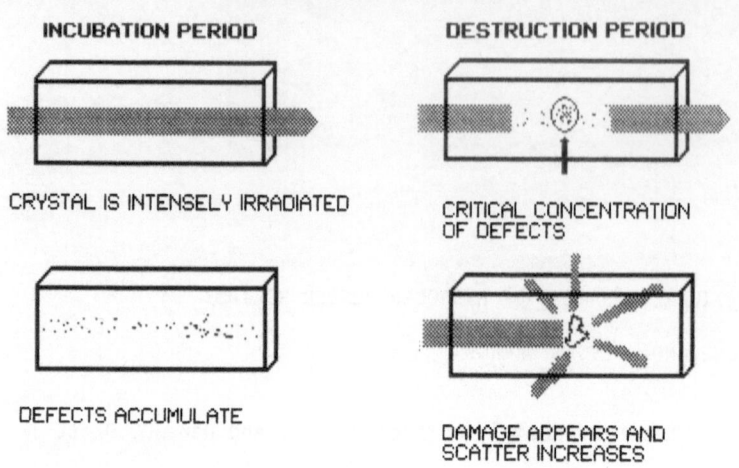

Figure 1. Schematic of an accumulated damage process.

The ionization process was characterized by "damage-detected spectroscopy". In this case multicolor picosecond pulse sequences were used to produce damage in many identical regions of a single crystal. Those sequences which pumped the excited singlet state via two-photon absorption, and then subsequently ionized this state via another two-photon absorption promoted maximum damage. "Damage-detected spectroscopy" is in a sense a solid-state analog of multiphoton ionization detected spectroscopy which is often used to study gas phase intermolecular dynamics.

The defect accumulation is apparently difficult to detect via conventional absorption or emission spectroscopy, but we have succeeded in detecting the accumulation with psCARS. We made the observation that while crystals were incubating, the long-lived optical phonons showed continuously decreasing lifetimes. Using impurity doping experiments and a calculation of the scattering rate of phonons from point defects, we have shown that the observed lifetime decrease is consistent with a critical concentration which is quite large [3].

The large defect concentration prior to crystal damage leads us to speculate that the transition from incubation to destruction occurs when the defects begin to interact with each other in photon assisted processes. This is quite different from the mode of accumulated damage in inorganic glasses, which is believed to be due to avalanche breakdown occurring in the vicinity of a very few defect centers [4,5]. An important question which needs to be answered is how a continuous process such as the accumulation of defects leads to such a sudden transition to a destructing state. Percolation [5] seems to be a natural model for such a process.

A variety of defect states can be created by the ~9eV ionization process. These include voids, self-trapped ion pairs, or diradicals. Apparently very energetic species combine in these crystals to produce the optical damage. In some sense this process seems to be a "solid-state explosion". The explosion is produced by interaction between energetic chemical species produced by laser pumping and is confined to a microscopic volume of roughly 25 microns in diameter within the cold crystal.

We have been able to estimate the irreversible work performed on the crystal necessary to form a spherical inclusion in the crystal [3]. This work is roughly 20μJ in a volume of 10^{-8} cm^3, corresponding to (translational) energy densities of 2,000 J/cm^3. By comparison, the heat of combustion of TNT is 22,000 J/cm^3. However it is known that for chemical explosives, only 10% of the heat of reaction is converted into translational energy, suggesting that the microscopic solid-state explosion is roughly as energetic as TNT and could serve as a model system to simulate explosive behavior.

We have succeeded in time-resolving the solid state explosion by detecting the transmission of 85 picosecond damage pulses through the sample. Two such damage runs are shown in Fig. 2. This figure plots the optical density of the sample, as determined by the integrated transmission of single picosecond pulses, versus the irradiation time. This time base is determined by treating the pulses as rectangular with 85ps width.

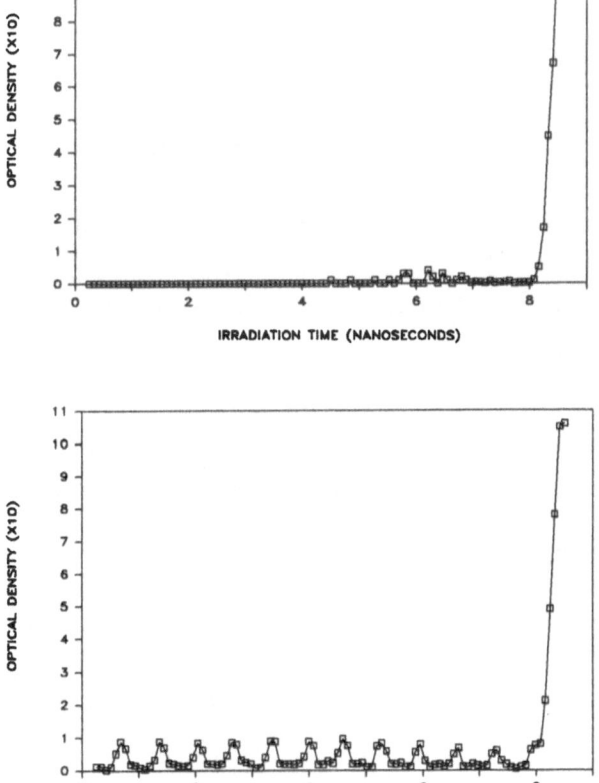

Figure 2. Time-resolved damage events showing the fast damage mode, and an oscillating damage event. Several metastable crystal states are involved in these processes.

In the upper data set, the optical density is constant for thousands of laser shots. About 2 ns before the catastrophic damage, small fluctuations are seen. The growth of the damaged volume occurs in less than 1 ns of irradiation. In the lower data set, well-defined oscillations in the optical density are observed. This oscillating optical density phenomenon is observed on about 10-20% of the damaged runs. In this run, the optical density was unchanged from 0 to 600 ns. At this point, the oscillations began abruptly. The data set shows only the time between about 745 and 755 ns, at which point the destruction commenced.

Oscillating light-pumped reactions are not new, but this is, to our knowledge, the first observation of such a phenomenon in the solid state. These reactions are usually described by a feedback scheme between the laser light absorption, the temperature of the system, and the temperature dependence of the equilibrium constant [6]. In our system these oscillations occur faster than phonon relaxation rates, so an analogous treatment would involve feedback with nonequilibrium phonon populations.

In the oscillating data set six distinct crystal states (five are metastable) are observed. These are 1) equilibrium low temperature crystal, 2) crystal with accumulated defects, 3) metastable transparent state (the state in the valleys of the oscillations), 4) "up-state", (in this state, the next laser pulse increases the OD) 5) "down-state" (next pulses reduces the OD) and 6) destructing state (subsequent pulses lead to catastrophically increasing damage). We are currently attempting to understand the nature of these metastable crystal states. One promising approach treats the system as a chemical relaxation oscillator [7]. We presume the existence of phonon and photon assisted processes which create and anhilate the defects to release the stored energy. Ultimately the defect concentration reaches its steady state value. However, in a relaxation oscillator, deviations from this steady state can cause the defect concentration to oscillate, or increase explosively, depending on the damping.

In this model, it is the fluctuations in the defect creation rate which ultimately determine the fate of the crystal, by driving it into oscillating or destructing states. These fluctuations are due to laser power fluctuations amplified by the multiphoton nature of the defect creation process. The sensitivity of a relaxation oscillator to the pumping process raises the intriguing idea that the behavior of the damaging crystal might be drastically altered by modulating the input laser beam [6].

A significant portion of this conference was devoted to hole-burning studies, particularly nonphotochemical hole burning (NPHB). In NPHB, a chromophore molecule absorbs light and then transfers energy to its surroundings. The local environment is then changed in some still uncharacterized manner which frequently shifts the chromophore [8]. This process is most facile in hydrogen bonded solids [8]. It may be viewed as caused by a local "melting". The type of process we observe may be related to NPHB, however it is certainly more violent. Figure 3 shows that so called "hole-burning" experiments are not aptly named--hole burning is a more appropriate description for our optical damage experiments!

ACKNOWLEDGEMENT

This research was funded by the National Science Foundation, Division of Materials Research. Dana D. Dlott acknowledges support from an Alfred P. Sloan Foundation fellowship.

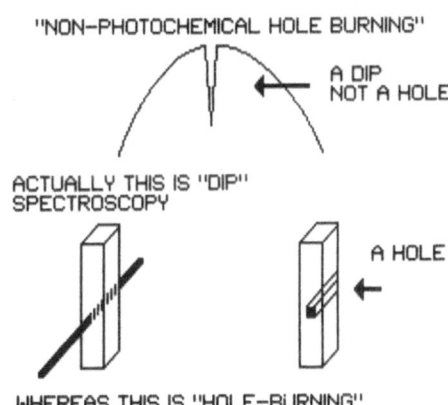

"NON-PHOTOCHEMICAL HOLE BURNING"

A DIP
NOT A HOLE

ACTUALLY THIS IS "DIP" SPECTROSCOPY

A HOLE

WHEREAS THIS IS "HOLE-BURNING"

Figure 3. "Hole-burning" experiments which were described at this conference merely produce a dip in the absorption spectrum whereas optical damage experiments burn a real hole!

REFERENCES

* Present address: Hughes Aircraft Corporation, El Segundo, California.

1. Thomas J. Kosic, Raymond E. Cline, Jr. and Dana D. Dlott, Chem. Phys. Lett. 103, 109 (1983).
2. Thomas J. Kosic, Raymond E. Cline, Jr. and Dana D. Dlott, Chem. Phys. 81, 4932 (1984).
3. Jeffrey R. Hill, Thomas J. Kosic, Eric L. Chronister, and Dana D. Dlott in Time-Resolved Vibrational Spectroscopy v. II (Lauberuau and Stockberger, Eds. 1985, in press); Thomas J. Kosic, Jeffrey R. Hill and Dana D. Dlott. Chemical Physics, (accepted 1985).
4. For a recent review of the optical damage field see (a) H. E. Bennett, A. H. Guenther, D. Milam and B. E. Newman, Ap. Opt. 22, 3276 (1983), or (b) Optical Engineering v. 22, no. 4 (1983).
5. J. M. Ziman, Models of Disorder (Cambridge University Press, 1979).
6. E.C. Zimmermann, M. Schell, and J. Ross, J. Chem. Phys. 81, 1327 (1984).
7. Oscillations and Traveling Waves in Chemical Systems, Field and Burger Eds. (Wiley, 1985).
8. G. J. Small, in Spectroscopy and Excitation Dynamics of Condensed Molecular Systems, V. M. Agranovich and R. M. Hochstrasser eds. (North Holland, 1983).

AN INTEGRATED FRACTAL ANALYSIS OF SILICA:

ELECTRONIC ENERGY TRANSFER, SMALL-ANGLE X-RAY-SCATTERING AND ADSORPTION

D. Rojanski-Pines (a), D. Huppert (a), H.B. Bale (b),
X. Dacai (b), P.W. Schmidt (b) D. Farin (c),
A. Seri-Levy (c) and D. Avnir (c)

(a) School of Chemistry, Tel Aviv University, Tel Aviv,
69978 Israel; (b) Physics Dept., University of Missouri,
Columbia, Missouri 65211; (c) Institute of Chemistry,
The Hebrew University of Jerusalem, Jerusalem 91904, Israel

ABSTRACT

The fractal dimension, D, of mesoporous silica gel was determined by
four independent approaches: Analysis of one step dipolar energy
transfer between adsorbed rhodamime B and malachite green; Analysis of
small-angle X-ray scattering at the Porod regime; tiling the surface
with molecules of different cross sectional area by adsorption;
Analysing the change in surface area as a function of particle size. All
four techniques agree in pointing to extreme surface wiggliness with
$D \to 3$. Besides confirmation of earlier indications to this high value,
our multiple approach contributes to the mutual corroboration of the
various theories involved. For comparison, flat-surfaced non-porous
silica (Aerosil) was studied. All methods give $D = 2.0$.

INTRODUCTION

Self similarity, or invariance to scale transformation, is quite
common in a wide variety of objects and processes. Recognition in this
phenomenon evolved rapidly in recent years, due mainly to Mandelbrot's
formulation of Fractal Geometry. In virtually all domains of the natural
sciences attempts have been made, in many cases with positive results,
to describe complex geometric environments and processes in such
environments in terms of fractal dimensions. It is beyond the scope of
this report to review these studies. However, in general one finds that
experiments in this field lag behind theory and simulations. It has been
the intention of this study to test experimentally various independent
theories on the same fractal object.

Common to all these theories is the general strategy in detecting
fractality: performing a resolution analysis, i.e. analysing the
decrease in a measurable property, as the yardstick size increases.
Thus, in an electronic energy transfer (ET) experiment, one studies the
decrease in ET rate as a function of donor-acceptor separation [1]: in a
small angle experiment one studies the decrease in amount of light which

passes a sample as a function of the size of the scattering unit [2,3.]: and in an adsorption study one analyses the decrease in apparent surface area as a function of the size of a probe molecule [4]. We describe now the application of these approaches to mesoporous silica gel [5].

RESULTS AND DISCUSSION

In a series of adsorption experiments and analyses of published adsorption data it has been obtained repeatedly that D of silica surface is close to the maximal value 3 [6]. This was determined by two types of analyses [7]:

$$n \sim \sigma^{D/2} \tag{1}$$

$$n \sim d^{D-3} \tag{2}$$

In the first, one determines the number, n, (moles/gr adsorbent) of molecules of cross sectional area, o, necessary to form a monolayer. In the second method, the yardstick molecule is kept constant, but the size of the object (particle diameter d) is changed. Both methods have yielded D values in the range 2.94 - 3.05 $\pm$ 0.05 [6], which is the highest possible fractal dimension for a surface. We add now some new evidence. Silica [5] was sieved to the fractions indicated in Table 1. For each fraction, monolayer values were obtained by the N_2 - BET method [8]. From eq. (2), D = 3.00 $\pm$ 0.01. The range of surface self-similarity [7] is 16 - 114 $\overset{\circ}{A}{}^2$. It is interesting to notice that geometrical considerations dominate the absorption behaviour, much more than the details of the adsorbent-adsorbate physisorption interactions. Thus, to an analysis of alcohols adsorption from solution (reported in detail in ref. 6a) according to eq. (1), we add now another experimental point (Fig.1) obtained by the completely different N_2 - BET method. The agreement is very good: D = 2.97 $\pm$ 0.02 without N_2, D = 3.06 $\pm$ 0.05 with N_2. Results of Drake et al suggest [9] that when chemical reactions are involved in these systems (chemisorption), geometrical consideration become, as expected, insufficient.

Table 1. Adsorption Data

Average particle diameter (microns)	Surface area (sq.m/gr)*
71.5	498
90	493
112.5	516
132.5	489
150	496
170	490
190	501

(*) nitrogen-BET method

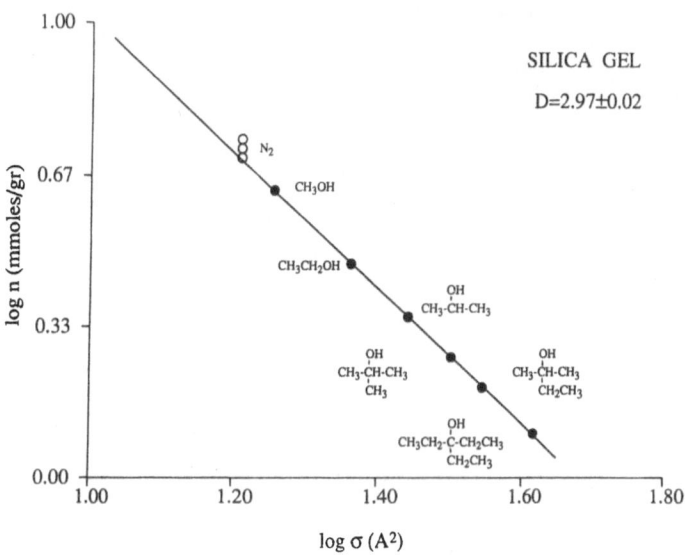

Figure 1. Resolution analysis, by adsorption. Monolayer values as a function of cross sectionas areas of adsorbed alcohols and adsorbed nitrogen on silica.

In a second family of experiments, Forster type electronic energy transfer (ET) between adsorbed rhodamine B (the excited donor, RB), to absorbed malachite green (the acceptor, MG) were studied on the silica. The application of ET for detection of fractal properties of materials was suggested by Klafter and Blumen [1,10]. The experimental method rests on time resolved measurements of direct long-range singlet inter-molecular ET from a donor molecule to an ensemble of acceptors randomly distributed on a fractal object. The survival probability P (t) of the excited donor is given by [10]:

$$P(t) = \exp(-P_1(t/\tau)^{D/6} - t/\tau) \tag{3}$$

$$P_1 \sim X_A \tag{4}$$

where is radiative life time of the donor in the presence of acceptors who occupy a fraction $_A$ of adsorption sites ($X_D \ll X_A \ll 1$). (For the two cases described here, (D 3 and D = 2.0), separate notations for surface and mass fractal dimensions are unnecessary). The silica was coated as previously [11] with RB or RB+MG at concentrations in the ranges 2×10^{-8}-5×10^{-8} moles RB/gr silica and 1.5×10^{-7}-6×10^{-7} moles MG/gr.

In order to investigate the ET process, the samples were excited with the 2nd harmonic pulse of a Nd^{3+} / YAG laser (25 psec FWHM, 1 Hz repetition rate). The fluorescence was imaged onto the entrance slit of a C939 Hamamatsu streak camera whose output was recorded and digitized by means of a PAR 1205D optical multichannel analyzer interfaced to a microcomputer for data processing. Each fluorescence curve represents the signal average of 30 separate laser shots. The experimental setup has already been described elsewhere [12]. The fluorescence of RB was separated from residual MG by means of optical filters. The time resol-ved decay of RB was simulated by the convolution of the excitation pulse and the function $\exp(-t/\tau)$. It was found that the decay profile for RB on the silica is a simple exponential with τ_o = 4.8 ± 0.1 nsec. This value is somewhat higher than $\tau_0 = 2.5$ns found in methanol. The increase in life time is probably due to inner-field effects [13] exerted by the adsorbate-adsorbent interactions, and not due to increase in fluorescence quantum yield which is already maximal (0.9 - 1) in solution.

The decay of the RB + MG system was simulated by the convolution of the survival probability (Eq.3) and the excitation pulse. With use of the value of RB decay the best value of P_1 and D were extracted by a mean least squares fit analysis, for a number of χ_A coverages (Table 2 and Fig.2). The resulting fractal dimension, $D = 3.0 \pm 0.3$ persists from Dexter distance ($\sim 5Å^2$) up to at least Forster radius ($\sim 100Å$). P_1 changes linearly with χ_A as predicted by eq.(4) [10]. These parameters are insensitive to χ_D within the range indicated above.

In a third set of experiments small angle X-ray scattering data was analysed. Analysis of scattering of various parts of the electromagnetic spectrum and of projectiles emerged recently as a powerful tool in structural characterization of irregular objects [2,3,14]. The scattering data were recorded with a Kratky camera [15]. A copper-target stationary-anode tube was the source of the X-rays, which had a wavelength 1.54 Å. For the scattering measurements, the dried silica sample was put in a stainless steel sample cell with Mylar windows. The scattered X-rays were detected with a position-sensitive detector [16]. Data for the inner and outer parts of the scattering curve (Fig.3) were obtained at two different detector positions with suffcient overlap between the two parts. After substraction of background scattering the curve was corrected for the effects of width [17] and height [18] of the collimating slits. Fig.3 is the corrected scattering curve.

We have shown [2,19] that for pores with fractral boundary surfaces, the intensity of scattered irradiation, I, is proportional to $q^{-(6-D)}$ where $q = 4\pi\lambda^{-1}\sin(\theta/2)$, λ is the X-ray wavelength and θ is the scattered angle. The slope of the outer part of Fig.3 is -4 rather than $-(6-3) = -3$, as one might expect from the experiments described above. However, the pre-factor in the power-law in eq. (6) of ref. 2 becomes zero for $D = 3$ resulting in Iq^{-4} and so our previously developed equation cannot afford a clear-cut distinction between $D = 2$ and $D = 3$. Although one can use the common knowledge that the surface of mesoporous silica gel is not flat [20] in order to exclude the $D = 2$ possiblity (see also comparison to non-porous silica below) we apply here a modified scattering equation aimed at reducing this problem.

Table 2.   Energy Transfer Data

Acceptor concentration* (mole/gr)	P_1	D
3.5×10^{-7} (a)**	2.0	3.0
7×10^{-7} (b)	3.8	3.0
14×10^{-7} (c)	5.7	2.9

(*) acceptor concentrations were determined by spectroscopic methods
(**) a, b, c in Fig.2.

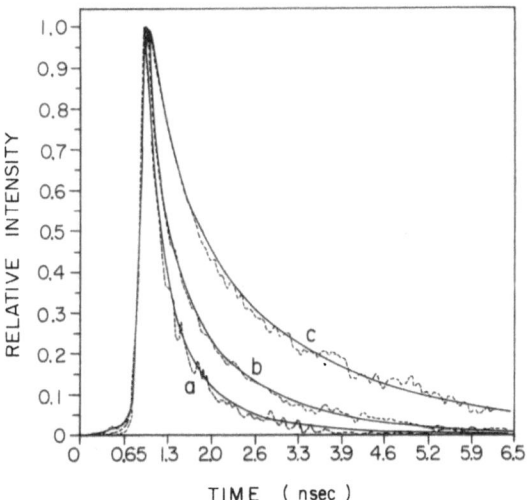

TIME (nsec)

Figure 2. Resolution analysis by an energy transfer experiment. Emission
intensity of rhodamine B in the presence of three different
concentrations (see Table 2) of malachite green on silica, as
a function of time. Full line: simulation; broken line:
experiment.

To obtain the modified scattering equation needed for the analysis
of scattering curves from samples with pore boundaries for which D = 3,
we began with the expression [21]:

$$I(q) = Ioo \int_0^x g(r) \frac{\sin qr}{qr} \, dr \tag{5}$$

where $q = 4 \pi \lambda^{-1} \sin(\frac{\theta}{2})$; I (q) is the scattered intensity; I_{oo} is a
constant; and g(r) is the pair correlation function. (For the small
scattering angles at which we recorded our data, $q = 2 \pi \theta_0 / \lambda$). Following
a suggestion of S.K. Sinha, we multplied the small r approximation for
g(r) obtained in reference 2 by the factor $e^{-(r/\zeta)}$, where ζ is a length
with the order of magnitude of the large dimensions of the pores. After
evaluating the Fourier transform of this approximate pair correlation
function, we added constant B to the intensity to allow for the nearly
constant scattering produced by fluctuations with dimensions no greater
than a few Angstroms. The resulting expression for the scattered
intensity I (q) is

$$I(q) = \pi I_e \delta^2 \left(\frac{8c(1-c)V\zeta^3 No \, \Gamma(5-D)\zeta^{5-D} \sin((5-D)\tan^{-1}q\zeta)}{(1+q^2\zeta^2)^2 \quad q(1+q^2\zeta^2)^{(5-D)/2}} \right) + B \tag{6}$$

where I_e is the intensity scattered by a single election; δ is the
electron density of the silica (which is assumed to be constant, with
the pores being considered to be empty); V is the volume of the scattering sample; c is the fraction of this volume occupied by the pores; N_o
is a constant characteristic of the pores boundary; and $\Gamma(x)$ is the
gamma function. For $q\zeta \gg 1$. I (q) is proportional to $q^{-(6-d)}$ and to

175

q^{-4} when $D \neq 3$ and $D = 3$, respectively. Thus, according to Eq. (6), the scattering data in Fig. 3 are consistent with the adsorption data and ET data, which indicate that D is very near 3.

The result of an almost three dimensional surface for mesoporous silica gel is interesting because it demonstrates that surfaces can be so wiggly and so irregular that they approach space-filling properties. This is especially emphasized by comparison to the other extreme, which is the "classical" two-dimensional case. According to fractal geometry, a two dimensional surface is smooth and feature-less. Indeed, for Aerosil, which is non porous and has a smooth surface the result D = 2 is obtained by all three methods. In this type of silica, very small spherical beads of diameter 125Å are arranged in a low co-ordination number as long chain-like aggregates [22,23]. Consequently, on scales which are below typical wiggles (< 100Å) the surface is locally smooth. On such scales nitrogen adsorption data gave, by using eq.(2), D = 2.02 $\pm$ 0.06 [24], SAXS measurements suggest the same value on scales < 100Å [25]; and finally, ET experiments, similar to the ones described above, yield D = 2.0 $\pm$ 0.2 [26]. A low D value (essentially 2 [27]) was obtained by this method for porous Vycor glass [28]. Features of regularity in this object, larger than the ET scale [26] (from small angle neutron scattering, porosimetry [26,29] and microscopy [30] indicate that fractal geometry need not be the best tool for analysis of this case, as suggested recently also in model simulations for Vycor-like cases [3]. Furthermore, as pointed out to us by Van-Damme, Vycor porous glass is an equilibrium structure (it is formed by leaching out a phase-separated boron oxide), and as such is not expected to be fractal. Finally, two dimensional ET has been treated already in previous studies, not from a fractal point of view [32].

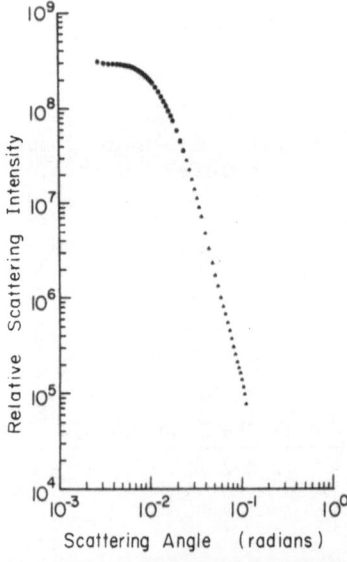

Figure 3. Resolution analysis by a SAXS experiment. The corrected small-angle X-ray scattering curve for the silica sample. The inner and outer parts of the curve were measured separately and are shown respectively by circles and triangles.

In conclusion, a multiple-technique approach seems to us important at this early stage of evaluation of theories and experiments which make use of the fractal concept. One should take into account that a possible healthy conclusion which may emerge from such studies could be that not everything under the sun is fractal. The recognition that valuable information can be extracted from resolution analyses in general, regardless of whether the result is fractal or not, is perhaps the first outcome of the recent intensive activity in this field.

ACKNOWLEDGEMENTS

Helpful discussions with many colleagues, especially with P.Pfeifer, S.K. Sinha and H. Van-Damme are deeply appreciated. Supported by grants from Israel Academy of Sciences, the US-Israel Binational Fund and the Szald Foundation. Assisted by the F. Haber Center for Molecular Dynamics, Jerusalem.

REFERENCES

(1) J. Klafter, A. Blumen and G. Zumofen, J. Lumin, 31/32: 627 (1984)
(2) H.D. Bale and P.W. Schmidt, Phys. Rev. Lett. 53: 596 (1984)
(3) D.W. Schaefer, J.E. Martin, P. Wiltzius and D.S. Cannel, Phys. Rev. Lett. 52: 2371 (1984)
(4) D. Avnir, D. Farin and P. Pfeifer, Nature (London), 308: 261 (1984)
(5) Silica, Woelm, No.04662; Lot No.90754/B; Particle size 63-200 m; N_2-BET surface area 500-600 m^2/gr average pore diameter 60Å.
(6) (a) D. Farin, A. Volpert and D. Avnir, J.Am. Chem. Soc.107:3368 (1985); (b) D. Avnir and P. Pfeifer, Nouv. J. Chim; 7.71 (1983) (c) D. Avnir, D. Farin and P. Pfeifer, J. Colloid Interface Sci. 103:112 (1985).
(7) P. Pfeifer and D. Avnir, J. Chem. Phys. 79: 3558 (1983)
(8) S. Brunauer, P.H. Emmett and E. Teller, J. Am. Chem. Soc. 60: 309, (1938).
(9) M. Drake et al, presented at this symposium.
(10) J. Klafter and A. Blumen, J. Chem. Phys. 80: 875 (1984)
(11) D. Avnir, R. Busse, M. Ottolenghi, E. Wellner and K. Zachariasse, J. Phys. Chem. 89: 3521 (1985).
(12) D. Huppert, H. Kanety and E. M. Kosower, Chem. Phys. Lett. 84:48, (1981).
(13) A.B. Myers and R.R. Birge, J. Chem. Phys. 73: 5314 (1980)
(14) E.g. D.L. Jordan, R.C. Hollins and E. Jakeman, Appl. Phys. B31:179, (1983).
(15) O. Kratky and Z. Skala, Elektrochem, Ber.Bunsens. Phys. Chem. 62: 73 (1958).
(16) Technology for Energy Corp. (Knoxville, Tennessee) Model 210
(17) T.R. Taylor and P.W. Schmidt, Acta Physica Austriaca 25: 293 (1967)
(18) J.S. Lin and P.W. Schmidt, J. Appl. Cryst. 7:439 (1974)
(19) P.W. Schmidt and X. Dacai, submitted.
(20) (a) K.K. Unger, "Porous Silica" J. Chromat. Library, Vol.16, Elsevier, Amsterdam, 1979; (b) R.K. Iler, "The Chemistry of Silica" John Wiley & Sons, New York 1979.
(21) A. Guinier, G. Fournet, C.B. Walker, Jr. and K.L. Yudowitch, "Small-Angle Scattering of X-rays", Wiley, New York, 1955, pp.78-79 (Equations (106) and (107).
(22) P.J.M. Carrot and K.S.W. Sing, Adsorption Sci.Tech., 1:31 (1984)
(23) P. Pfeifer. Appl. Surf. Sci. 18: 146 (1984)

(24) D. Avnir, D. Farin and P. Pfeifer, J. Chem. Phys. 79: 3566 (1983)

(25) D.W. Schaefer, J.E. Martin, A.J. Hurd and K.D. Keefer, preprint 1984

(26) D. Pines-Rojansky, D. Huppert, D. Avnir, Chem. Phys. Lett. 139:109 (1987).

(27) R. Kopelman, Book of Abstracts of this Symposium.

(28) U. Even, K. Rademann, J. Jortner, N. Manor and R. Reisfeld, Phys. Rev. Lett. 52: 2164 (1984). Also in J. Lumin, 31/32: 634 (1984)

(29) B.V. Enustun, J. Eckrich and T. Demirel, Proc. Int. Symp. Particulate and Multi-Phase processes, Florida Apr.1985.

(30) K. Kadokura, Ph.D. Dissertation, University of California, Los Angeles, 1983.

(31) C.L. Yang, P. Evesque and M.A. El-Sayed, J. Phys. Chem. 89: 3442 (1985).

(32) A.G. Tweet, W.D. Bellamy and G.L.Gaines, J. Chem. Phys. 41: 2068 (1964); M. Hauser, U.K.A. Klein and U. Gosele, Z. Phys. Chem. N.F. 101: 255, (1976); P.K. Woller and B.S. Hudson, Biophys J. 28:197 "Fluorescent Probes", G.S. Beddard and M.A. West, Ed's, Academic Press, London 1981.

STRUCTURAL CHANGES AND OSCILLATIONS DURING THE SOL-GEL TRANSITION IN SILICA GEL GLASS AS PROBED BY PYRENE EXCIMERIZATION

Vered R. Kaufman and David Avnir

Department of Organic Chemistry
The Hebrew University of Jerusalem
Jerusalem, 91904, Israel

INTRODUCTION

Polymerization of metal alkoxides as a method for low temperature synthesis of glass is a rapidly developing art, due to the significant advantages it has over the classical high-temperature melting technique [1]. Considerable attention has been given mainly to the transition of the dry porous gel to the monolithic non-porous glass [2]. We have decided to concentrate on the early stages of the glass formation i.e., on the sol to dry gel transition. This transition which has been investigated by several groups [1,2] determines the properties of the final glass; hence the importance of its investigation. In a recent series of papers, we have introduced a sensetive analytical tool for the study of the sol to gel to glass transitions: photoprocesses of probe molecules whioh are embedded in the reacting system [3-6]. This analytical tool which is well developed for biological systems, and finds increasing use in surface science [7-9], is based on the idea of using photoprobes which are highly sensitive to environmental parameters such as polarity, viscosity, surface adsorption sites, degree of porosity and degree of geometrical-complexity (fractal dimension). Here we report on complex structural changes, including periodicity, that occur during the sol-gel transition, as probed by pyrene photophysics.

PYRENE AS A FLUORESCENT PROBE FOR THE SOL-GEL TRANSITIONS

Pyrene has been used quite intensively as a probe for heterogeneous systems [7,8]. The long singlet life-time, the well resolved vibronic structure and the intermolecular comlexation process between an excited pyrene and a ground state pyrene (excimer formation, Fig. 1) have contributed to the wide use of this molecule. When pyrene is added to a standard silica sol-gel starting solution [4], gradual changes are observed in the emission spectrum, as the gelation and dehydration process proceeds (Fig. 1,2). The amount of the intermolecular product, the excimer, gradually increase, until a maximum is reached (Fig. 1D,2). From then on, the amount of excimer decreases until

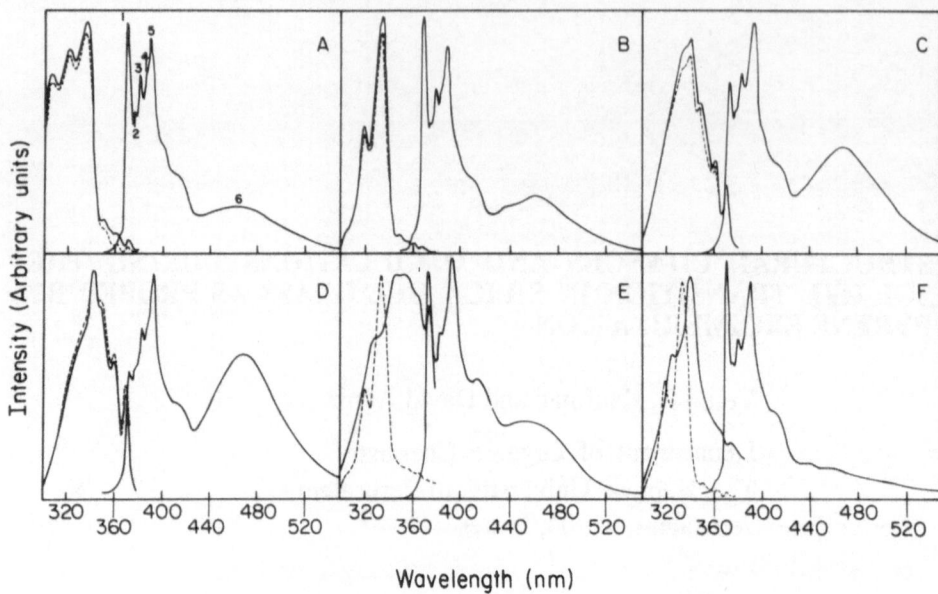

Figure 1. Emission (right) and excitation (left; λ_{ex} =470nm, and λ_{ex} =392nm) spectra of pyrene (1×10^{-3} M) at six stages of the sol-gel process. A. Starting solution, a typical monomer (peaks 1-5) and excimer (peak 6) emission. B. The gel point, characterized by the overlap of the excitation spectra. C. and D. Soft gel, characterized by growing of the excimer, decrease in the monomer, and by the broadening of the excitation emission. E. Transition from soft to hard gel, characterized by the decrease in the excimer emission and the separation of the monomer and excimer excitation emissions. F. Final hard gel. The excimer disappeared. Isolation of the pyrene molecules as monomers. Excitation spectrum of residual non-isolated molecules is shifted.

only residual excimer can be detected (Fig. 1F,2). The complete isolation [3,4,6,10] of the pyrene molecules at the final room-temperature drying stage is unique to the sol-gel process. Ground-state pyrene aggregation, which is revealed by excimer like emission and by spectral shifts of the excitation spectrum, is commonly observed on silica surfaces [7,8]. The gradual changes in Fig. 2 follow, step by step, the structural changes in the SiO_2 matrix: the intial hydrolysis forms non porous particles of silica having both Si-OH and Si-OR groups on the surface. On the smooth surface of these particles, the excimer is mainly dynamic and still not pronounced. As the particles aggregate, a loose gel of macroporous silica is formed, which slowly dries to microporous silica. During this transition, the amount of excimer slowly grows and it has an increase static excimer contribution. The material structure is that of silica gel and one gets the known behavior on silica surfaces. As the cages continue to form, more and more pyrene molecules are forced into the unstable repulsive intermolecular distance (4A [11]), and a molecular isolation process starts to operate. A different pattern of behavior is observed during the formation of reversed phase silica gel glasses, by hydrolysis of compounds of the type $Si(OR)_3R'$, because of the reduction in the cross-linking positions and because of the hydrophobic nature of the R' group (methyl, ethyl, phenyl and others) [6,12]. Typically, the amount of excimer in these systems remains low throughout the whole process [6,13]. As shown in Fig. 2, there are differences in the kinetic

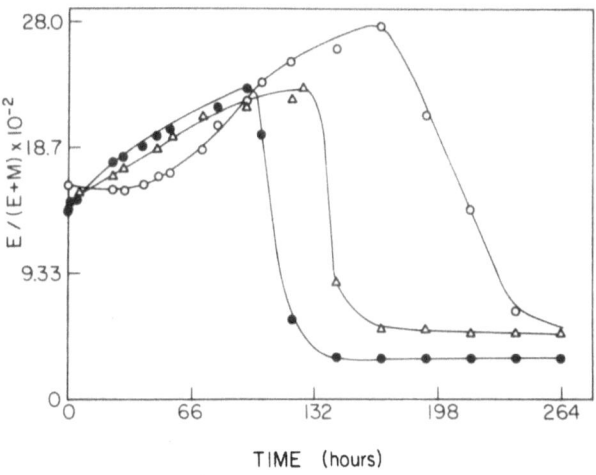

Figure 2. Effect of pH on the amount of excimer (E/E+M) as function of time. ●
pH=9.8, O pH=7.5 and Λ pH=2.3.

behavior, as a function of time and pH. Under basic conditions, the time required for reaching maximum excimer emission is the shortest. Under acidic conditions this time is somewhat longer, and the slowest process is under neutral conditions. These differences are direct reflection of the well known changes in the hydrolysis-condensation stage as a function of the pH [1,14]: Under basic conditions the more branched and cross-linked polymer is formed, resulting in relatively short time for gelation, whereas the acidic polymerization forms linear chains which require longer time to gel. No difference in reaction rate is found in the desication stage (right slope in Fig. 2), indicating perhaps, that most of the alkoxy groups are already hydrolyzed at this stage. As expected, under neutral conditions, the process is the slowest. Another interesting observation is that under neutral conditions, the maximum excimer emission is higher than under non-neutral conditions. This, again, may be a reflection of a faster caging process that occurs under the latter conditions. It may also indicate the smaller average pore size obtained under neutral and acidic conditions [14], causing, at first, higher amount of pyrene aggregation, followed later by slower equilibration (the right tail of the curve is less steep). A similar behavior was observed on silica with various pore size distributions [15].

PROLONGED OSCILLATONS DURING GEL AGING

There has been recently intensive interest in the evolution of spatial and temporal patterns (dissipative structures), in systems driven away from equilibrium [16]. Chemical oscillations are perhaps the most studied type of these structures [17]. Recently, we have reported on a unique oscillator, observed during the sol-gel process [6]. When surface active agent, (any of the following: Tweens, Spans, Brijs, Myrjs, Triton-X-100, ammonium salts and others), are added $(3.5 \times 10^{-5} - 1.5 \times 10^{-3}$ M) to the starting sol-gel solution [4], the system does not reach an equilibrium structure as shown in the right hand side of Fig. 2, but reveals an oscillatory behavior between the extreme possible values of the pyrene excimer emission intensities for at least 1200

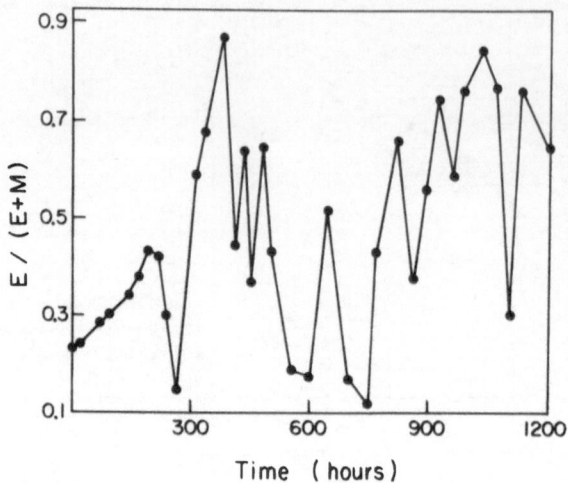

Figure 3. Oscillations in the relative fluorescence intensity of pyrene excimer as a function of time in a silica gel glass, prepared with a surface active agent (Triton-X-100, 3.5×10^{-5} M and 1×10^{-3} M pyrene). Oscillations are observed after about 240 hours, and continue for more than 1200 hours. For another example see Fig. 3 in ref. 6.

hours, with changing period of around several hours (Fig. 3). Much care has been taken to establish the authenticity of these oscillations e.g., by taking, for some of the oscillations, measurements every 10 minutes [13]. Remarkably, the oscillations are synchronous throughout the monolithic gel glass. Periodicities were observed also with other probes: pyrene butyric acid and pyrene carboxyaldehyde. When 1,3-dipyrenylpropane is used as a probe, no surface active agent is needed for observing the oscillations. The underlying mechanism of these oscillations is not completely clear at this stage. The oscillatory behavior may bear on the question of dynamic "breathing" and reversibility of the formation of the Si-O-Si bond [18]. Klonowski [19] has developed a theoretical model which predicts the possibility of oscillations between sol and gel. Oscillations in open reactor polymerizations have been studied both theoretically [20] and experimentally [21]. The relevance of all these reports to the phenomenon we found is under intensive study.

ACKNOWLEDGMENTS: Supported by the NCRD, Israel and by the KFA, Julich, West Germany. Assisted by the F. Haber Research Center for Molecular Dynamics. Helpful discussions with D. Levy and E. Wellner are acknowledged.

REFERENCES

[1]. a). S. Sakka and K. Kamiya, J. Non-Cryst. Solids 42:403 (1980); b). H. Dislich, J. Non-Cryst. Solids 57:371 (1983); c). C. J. Brinker and G. W. Scherer, J. Non-Cryst. Solids 70:301 (1985).
[2]. a). J. Zarzycki, M. Prassas and J. Phalippou, J. Mat. Sci. 17:3371 (1982); b). R. A. Assink and B. D. Kay, Mat. Res. Soc. Proc. 32:301 (1984); c). M. Nogami and Y. Moriya, J. Non-Cryst. Solids 37:191 (1980); d). M. Yamane, S. Inoue and A. Yasumori, J. Non-Cryst Solids 63:13 (1984); e). D. W. Schaefer, J. E. Martin, A. J. Hurd and K. D. Keefer, preprint 1985.
[3]. D. Avnir, V. R. Kaufman and R. Reisfeld, J. Non-Cryst. Solids 74:395 (1985).

[4]. D. Avnir, D. Levy and R. Reisfeld, J. Phys. Chem. 88:5956 (1984).

[5]. D. Levy, R. Reisfeld and D. Avnir, Chem. Phys. Letts. 109:593 (1984).

[6]. V. R. Kaufman, D. Levy and D. Avnir, Proc. 3rd Intr. Workshop "Glasses and Glass Ceramics from Gels", Montpelier Sept. 1985, J. Non-Cryst. Soilds, 82:103 (1986).

[7]. E.g., P. de Mayo, L. V. Natarajan and W. R. Wave, ACS Symp. Ser. 278 (1985) 1, and references cited therein.

[8]. D. Avnir, R. Busse, M. Ottolenghi, E. Wellner and K. Zachariasse, J. Phys. Chem. 89:3521 (1985); P. Levitz H. Van-Damme and P. Keravis, J. Chem. Phys. 88:2228 (1984).

[9]. Z. Grauer D. Avnir and S. Yariv, Can. J. Chem. 62:1889 (1984); D. Rojanski et. al., this volume; D. Levy and D. Avnir, this volume; A. Levy, D. Avnir and M. Ottolenghi, Chem. Phys. Lett. 121:233 (1985).

[10]. T. Tani, H. Namikawa, K. Arai and A. Makishima, J. Appl. Phys. in press (1985).

[11]. N. J. Turro, "modern Molecular Photochemistry", Benjamin Cummings, California, 1978.

[12]. H. Scholze, J. Non-Cryst. Solids 73:669 (1985); H. Schmidt, J. Non-Cryst. Solids 73:681 (1985).

[13]. V. R. Kaufman and D. Avnir, Langmuir, 2:717 (1986).

[14]. M. Yamane, S. Aso and T. Sakaino, J. Mat. Sci. 13:865 (1978).

[15]. E. Wellner, M. Ottolenghi D. Huppert and D. Avnir, Langmuir 2:616 (1986).

[16]. G. Nicolis and I. Prigogine, "Self-Organization in Non-Equilbrium Systems", Wiley, N.Y. 1977; D. Avnir and M. Kagan, Nature 307:717 (1984).

[17]. R. Noyes, B. Bunsenges, Phys. Chem 84:295 (1980).

[18]. C. Hyver, J. Chem. Phys. 84:850 (1985).

[19]. W. Klonowski, J. Mat. Res. submitted; W. Klonowski, J. Appl Phys. 58:2883 (1985).

[20]. V. A. Kirillov and W. H. Ray, Chem. Eng. Sci. 33:1499 (1978); J. W. Hamer, T. A. Akramov, and W. H. Ray, Chem. Eng. Sci. 36:1897 (1981); C. Kiparissides, J. F. MacGregor and A. E. Hamielec, J. Appl. Polym. Sci. 23:401 (1979).

[21]. R. K. Greene, R. A. Conzalez and G. W. Poehlein in "Emulsion Polymerization", p.341. I. Pirma and J. L. Gordon, Eds., ACS Symposium Series 24, Washington, D.C. 1976.

ORGANIC MOLECULE, 1,4-DIHYDROXY-9,10-ANTHRAQUINONE, DOPED AMORPHOUS SILICA PREPARED BY SOL-GEL METHOD AND ITS STRUCTURES STUDIED BY PHOTOCHEMICAL HOLE BURNING

Toshiro Tani, Akio Makishima*, Akira Itani** and Uichi Itoh

Electrotechnical Laboratory
1-1-4 Umezono, Sakura-mura Niihari-gun Ibaraki 305, Japan

*National Institute for Research in Inorganic Materials
1-1 Namiki, Sakura-mura Niihari-gun Ibaraki 305, Japan

INTRODUCTION

Ever since the anomalous contribution to the specific heat of silica glass below 1 K[1] was reported, extensive experimental efforts have been devoted to understand the dynamical properties of disordered solids. Although the observed phenomena seem to be consistent with the theoretical model based on the existence of two-level system's(TLS's)[2], the microscopic nature of TLS's is not yet understood. In our attempt to gain further insight into the structures and dynamical properties of amorphous solids at low temperatures, we tried to introduce photosensitive functional "organic dye molecules" into the "inorganic silica glass matrix" and to probe the amorphous structure by the photochemical hole burning(PHB)[3] spectroscopy.

The present paper describes the preparation and structures of amorphous silica doped with organic dye molecules as a new hybrid organic-inorganic materials[4]. 1,4-dihydroxyanthraquinone (DAQ, quinizarin) was embedded in a matrix of amorphous silica (DAQ/a-SiO$_2$) by the alcoholate method[4,5].

As silica glass is a typical amorphous oxide, introduction of functional organic molecules even though they being thermally dissociative and to probe its structural change optically through the doped molecules must be attractive. It has never been reallized due to the high melting temperature of silica. However, adopting the sol-gel method using tetraethoxysilane, Si(OC$_2$H$_5$)$_4$, as a starting material for silica gel glass formation, we obtained the organic molecule-doped inorganic glassy materials as a molecular dispersion, which is stable at ambient. The hydrolysis for gelling was carried out at 60 C with HCl as a catalyst (pH=5). The acidic, or at least neutral, hydrolysis of Si(OC$_2$H$_5$)$_4$ is a necessary condition for the formation of monolithic silica gel[6]. Under the basic condition, the syneresis occures quite often and the gel takes micrograin-like structure.

ABSOPTION AND FLUORESCENCE OBSERVATIONS

The absorption and fluorescence spectra of DAQ/a-SiO$_2$ system in

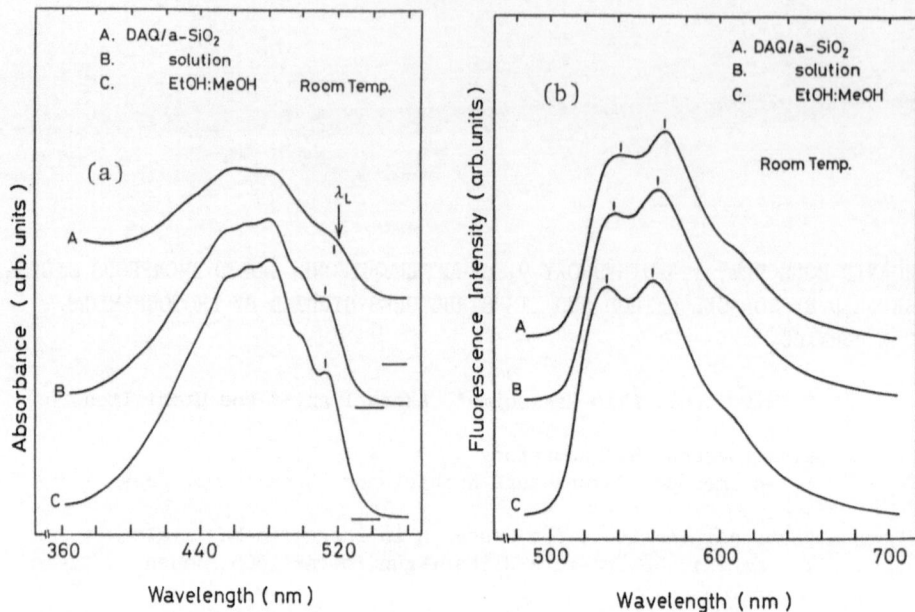

Fig.1   Absorption (a) and fluorescence (b) spectra of quinizarin at
room temperature.   A's in a-SiO$_2$, B's in Si(OC$_2$H$_5$)$_4$:ethanol:
water solution and C's in ethanol:methanol 3:1 mixed solvent[4] .

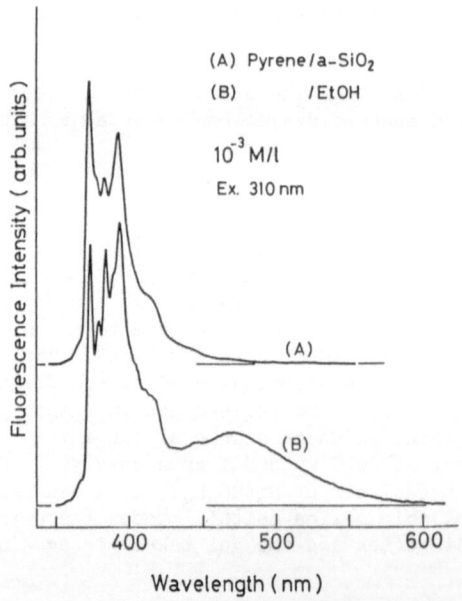

Fig.2   Fluorescence spectra of pyrenes in (A) a-SiO$_2$ and (B)
ethanol at room temperature.   Densities of pyrenes are
both 10^{-3} M/1 .

visible region measured at room temperature are shown by curve A in Figs.
1(a) and 1(b), respectively. The spectra of DAQ in the solution of
$Si(OC_2H_5)_4$/ethanol/water with HCl as a catalyst just after mixing (curve B)
and those in 3:1 mixed solvent of ethanol and methanol, EtOH–MeOH, (curve
C) are also shown for comparison.

The comparatively large background of curve A in Fig.1(a) is mainly
due to the surface scattering of the sample. As can be seen in the figures, the profiles of absorption and fluorescence spectra in $DAQ/a-SiO_2$
system resemble generally well those of DAQ in the liquid solutions. This
indicates that the quinizarin molecules are embedded in the bulk of amorphous silica without changing chemically. Moreover, assuming that the
oscillator strength of DAQ is the same in the solutions and in the rigid
matrix, the concentration of DAQ molecules in $a-SiO_2$ estimated from the
absorption spectrum is almost equal to that calculated from the starting
density simply by taking the volume contraction into account. This confirms further that the doped DAQ molecules are dispersed in $a-SiO_2$ without
cracking or decomposition.

One of the advantages of utilizing the rigid silica glass matrix is
the ability to suspend the dye molecules at extremely high dye concentration with molecular dispersion where dimerization or aggregation will occur
in the ordinary liquid solutions.

This can be seen mostly in pyrene/$a-SiO_2$ system as shown in Fig.2. At
the concentration of 10^{-3} M/l, broad excimer band is appeared around 470 nm
in the fluorescence spectrum in the ethanol solution. This excimer band
does not appear in the $a-SiO_2$ matrix at the same concentration of pyrene
molecules.

PHOTOCHEMICAL HOLE BURNING OBSERVATIONS

The PHB's were performed at low temperatures with a cw krypton ion
laser operated at 520.83 nm. The samples were cooled down to liquid helium
temperature region with the continuous flow temperature-variable cryostat.
The burned holes were detected at various temperature by measuring the
transmission spectra with the monochromated light from 150 W tungusten-
halogen light source through 1-m monochromator. A grating with a dispersion of 8 Å/mm and slit width of 10 μm give the resolution of about 0.08 Å.

Typical hole profile burned at 4.8 K under 0.74 mW/cm^2 for 30 minutes
irradiation is shown in Fig.3(a). The burning yield of zero-phonon hole is
1.2×10^{-4} and is almost equal to the one observed in the alcoholic glass.
Burning-time dependences of the holewidth at half depth are shown in Fig.4.
The holewidth increases almost linearly with burning time and saturation
broadening does not appear within the present experimental conditions. The
observed behavior can be interpreted by the equation expressed as follows;

$$\Gamma_h(t)= \Gamma_0(1+(\sigma I \phi \alpha /\hbar \omega_L)t)= \Gamma_0(1+Ct),$$

where σ is the absorption cross section, I and ω_L the laser light intensity and frequency, ϕ the photochemical yield and α the Debye–Waller
factor.

This model assumes phenomenologically a Lorentzian line shape of
molecular absorption irrespective of its line broadening mechanism. The
line shape of the hole is almost Lorentzian as can be seen from Fig.3(b).
From the data shown in Fig.4, the intrinsic holewidth Γ_0 extrapolated to
t=0 in this system is 0.9 cm^{-1} and the coefficient of time dependence is C
$= 1.7 \times 10^{-2}$ min^{-1} at 4.8 K. Assuming the absorption cross section is

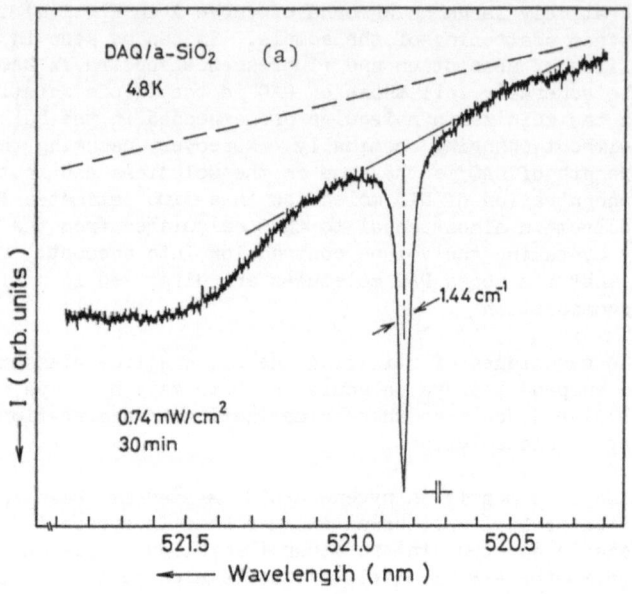

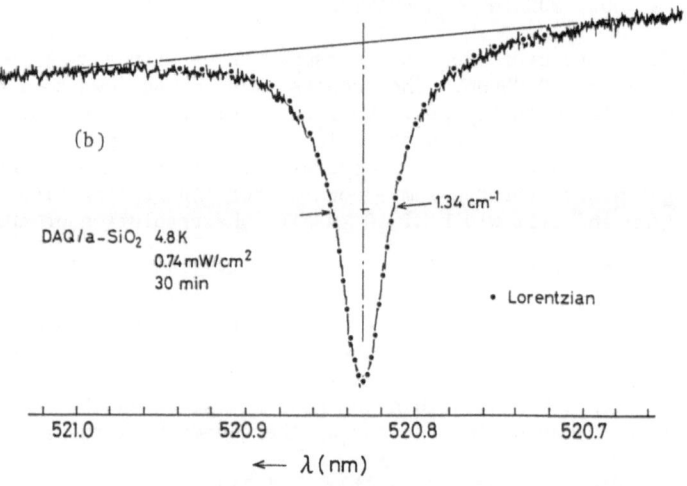

Fig.3 Hole spectra of quinizarine/a-SiO$_2$ burned at 4.8 K. Figure (a) shows zero-phonon hole at 520.83 nm in resonance with laser light and broad phonon-side hole around 521.5 nm[4]. Figure (b) shows the line shape of zero-phonon hole.

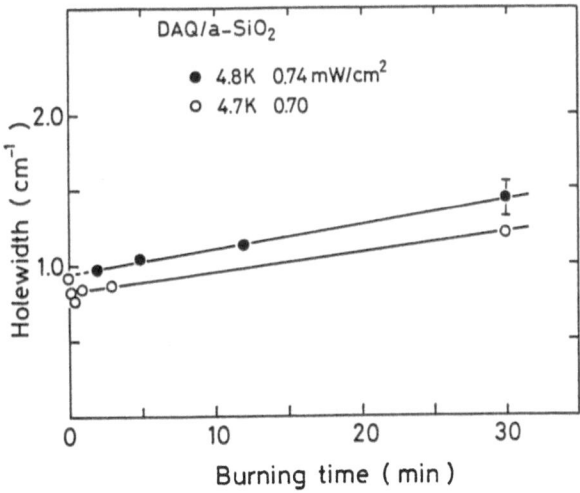

Fig.4   Burning-time dependences of zero-phonon holewidth[4].

common in the solutions and in a-SiO$_2$ matrix and does not change with temperature, the C value can be estimated experimentally as 2.3×10^{-2} min^{-1}, which agrees well to the observed value within our experimental accuracy.

The behavior of burning time dependence of DAQ/a-SiO$_2$ system is quite different from that observed in the alcoholic glass matrix. The C coefficient at short burning time region observed in the EtOH-MeOH matrix is more than one order of magnitude larger than in the a-SiO$_2$ if normalized by the laser intensity I. On the other hand, intrinsic holewidth Γ_0 is roughly the same in both cases; 1.1 cm^{-1} in EtOH-MeOH glass and 0.9 cm^{-1} in a-SiO$_2$. The holewidth Γ_0 refrects the optical dephasing process of the excited state which should be affected mostly by the nearest neighbour structure of the quinizarin molecules in the first approximation.

Therefore, the fact that Γ_0's in a-SiO$_2$ and EtOH-MeOH are almost equal to each other suggests that the nearest neighbour structure of quinizarin in a-SiO$_2$ is effectively the same as in the alcoholic glass: that is, hydroxylic nature of surroundings. Further, the fact that C coefficient in a-SiO$_2$ differs from that in EtOH-MeOH suggests that the quinizarin molecules are not enclosed by the ethanol and/or water molecules, but are almost directly surrounded by SiO$_2$ matrix.

Two kinds of silicon groups in a-SiO$_2$, silanols(Si-OH) and siloxanes (Si-O-Si), possibly contact with quinizarin directly. Our result, therefore, indicates that the surface of the cage is rich in silanol groups. Of course, we can not exclude the possibility of inclusion of one or a few ethanol molecules in a cage. As for water, we feel its possibility is rather low, because solubility of quinizarin is high in ethanol but rather low in water.

The annealing properties of the holewidth described below also serve as a support on this picture. The temperature dependences of hole profiles are most important for characterizing the DAQ/a-SiO$_2$ system. We adopted here two kind of annealing procedures. The hole was burnt at the base temperature(4.7-4.8 K). (i) The <u>stepwise annealing</u>: the temperature of the sample was raised in successive steps and the hole spectra were observed at each temperature. (ii) The <u>cycle annealing</u>: the temperature of the sample was raised to a prescribed point, being stayed for more than five minutes,

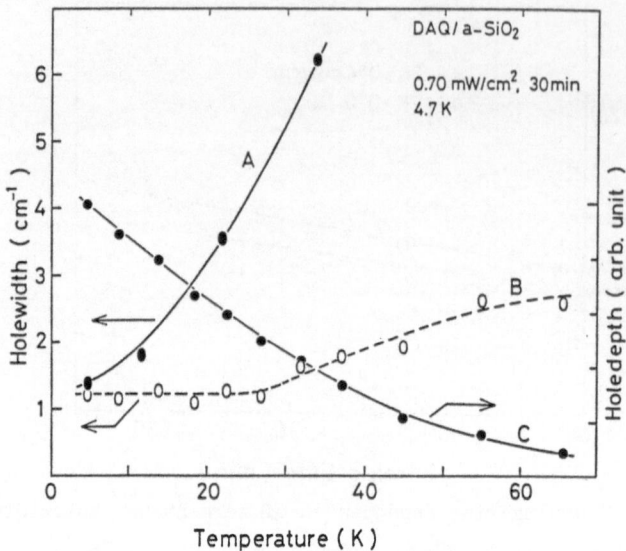

Fig.5   A: Temperature dependence of holewidth under the stepwise
annealing, B: Annealing profile of temperature cycling, C:
Change in the holedepth of B[4] .

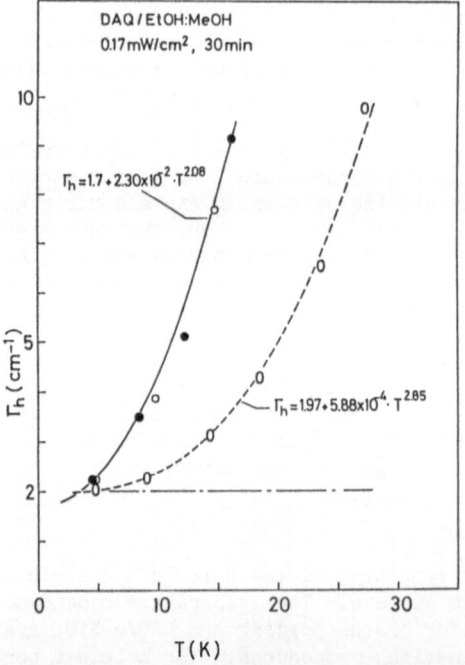

Fig.6   Temperature dependences of holewidth in DAQ/EtOH:MeOH system
observed under the same condition as in Fig.5. Solid curve
corresponds to stepwise annealing and dotted one to cycle
annealing.

then cooled down quickly to the base temperature. Hole spectra were all observed at base temperature.

The temperature change of the holewidth obtained by the stepwise annealing is shown in Fig.5-A, which is almost proportional to the square of the temperature. This may support the dephasing process through the interaction between the dye molecules and the TLS's inherent to the amorphous state. The T^2 behavior of homogeneous linewidth is predicted by Lyo and Orbach[7].

The most noticeable property in DAQ/a-SiO$_2$ system appears in the cycle annealing process. As shown in Fig.5-B, the hole survives even after the temperature cycling of as high as 67 K. The linewidth almost perfectly recovers under the temperature cycling up to 27 K. However, it begins to grow up at higher than 30 K or, in other words, the irreversible component appears and increases gradually with raising the cycle temperatures above 30 K. These behaviors are quite different from those observed in the EtOH-MeOH glass shown in Fig.6, where the irreversible change is large and increases steeply with cycle temperatures and the hole can not survive after cycle annealing above 30 K.

In the organic glasses made of small poler molecules such as EtOH-MeOH glass shown here, there seems to exist some cluster-like structure. It contains a lot of solvent molecules in somewhat ordered texture. These domains can be rearranged thermally in the temperature region of Fig.6, which consequently causes the irreversible spectral diffusion---hole broadening---in the organic glass systems. Detailed explanation on the organic glass systems will appear elsewhere[8]. In comparison with the organic glasses, the network of the present a-SiO$_2$ seems fairly rigid and contains no distinguished structures.

This anomaly appeares in the holewidth only and not in the holedepth as is shown in Fig.5-C. The hole filling backward reactions of photoproducts seem to be thermally induced monotonically. The anomaly

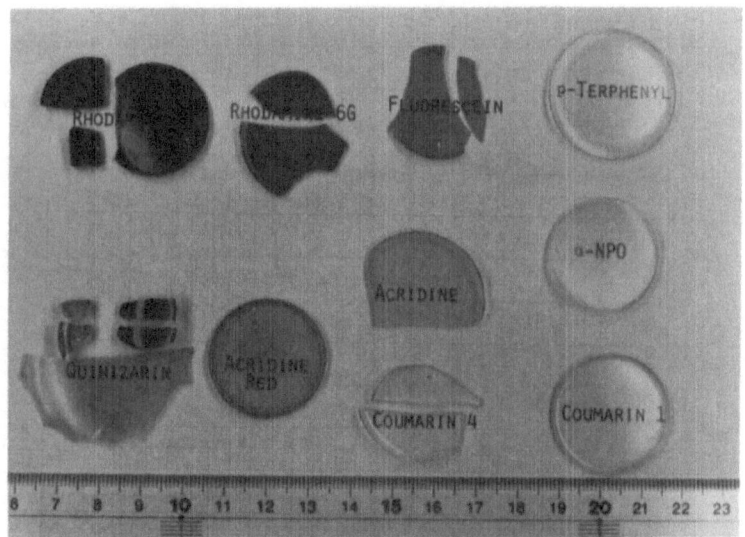

Fig.7   Several examples of the synthesized hybrid organic molecules/a-SiO$_2$ materials.

should be related to the relaxation processes which give rise to further broadening of the hole irreversibly due, for example, to the change in the ground state distribution of the sites. The microscopic origin of these processes is not yet clear. We speculate so far two possible mechanisms on the irreversible component appearing above 30 K cycle temperature. One is the rotational movement of oxygen atoms perpendicular to their bond direction (Si-O-Si) and the other is the residual organic groups such as $-OC_2H_5$ due to insufficient polymerization.

APPLICATIONAL ASPECTS

Adopting the sol-gel method, we can introduce various kinds of functional large organic molecules into the inorganic oxide glasses, which has a variety of compositions such as SiO_2, GeO_2, TiO_2, B_2O_3, P_2O_3 and their mixtures. In Fig.7, several examples of the synthesized hybrid organic molecules/a-SiO_2 materials are shown. These organic molecule doped inorganic glassy materials in the molecular dispersion may provide a new field in the material science. The materials will be promising also for numerous optical and optoelectronic applications in various ways including the PHB memory due to the advantage of its rigidity and stability of the inorganic glass matrices which keep alive the functions of organic molecules.

REFERENCES

1. R. C. Zeller and R. O. Pohl, Thermal conductivity and specific heat of noncrystalline solids, Phys. Rev. B4:2029 (1971).
2. P. W. Anderson, B. I. Halperin and C. M. Varma, Anomalous low-temperature thermal properties of glasses and spin glasses, Philos. Mag. 25:1 (1972),
   W. A. Phillips, Tunnelling states in amorphous solids, J. Low Temp. Phys. 7:351 (1972).
3. J. Friedrich, H. Wolfrum and D. Haarer, Photochemical holes: A spectral probe of the amorphous state in the optical domain, J. Chem. Phys. 77:2309 (1982).
4. T. Tani, H. Namikawa K. Arai and A. Makishima, Photochemical hole-burning study of 1,4-dihydroxyanthraquinone doped in amorphous silica prepared by alcoholate method, J. Appl. Phys. 58:3559 (1985).
5. A. Makishima and T. Tani, Preparation of amorphous silicas doped with organic molecules by the sol-gel process, J. Amer. Cer. Soc. to be published.
6. D. Avnir, D. Levy and R. Reisfeld, The nature of the silica cage as reflected by spectral changes and enhanced photostability of trapped rhodamine 6G, J. Phys. Chem. 88:5956 (1984).
7. S. K. Lyo and R. Orbach, Homogeneous fluorescence linewidths for amorphous hosts, Phys. Rev. B22:4223 (1980).
8. T. Tani, Matrix effects and structures of organic glasses made of small poler molecules studied by photochemical hole burning, under preparation.

** on leave from Mitsui Petrochemical Industries, Ltd. 2-5, Kasumigaseki 3, Chiyoda-ku Tokyo 100, Japan.

ISOLATION AND TRAPPING OF ORGANIC PHOTOACTIVE MOLECULES IN SILICA GLASS MATRICES AT ROOM TEMPERATURE

David Levy and David Avnir

Department of Organic Chemistry, The Hebrew University of

Jerusalem, Jerusalem 91904, Israel.

1. Preparation and Properties of the Novel Matrix

Matrix isolation is one of the principal tools in mechanistic studies of chemical reactions, especially of photochemical ones [1]. The two routine methods employed for reaching the situation of molecular trapping and isolation in solid environments are either cooling a mixture of organic solvents below its glass-forming temperature[1], or preparing a rigid polymeric plastic material from organic monomers [2].

In a recent series of papers [3-7] we have described the development and uses of another general method of matrix isolation which solves an old and notorious problem: How to embed an inorganic oxide glass with organic molecules. Standard glass preparation techniques cannot, of course, be used for these purposes because of the high temperatures of glass melting under which organic molecules will immediately burn. Consequently, our idea was to look for possible methods of preparing these inorganic glasses at much lower temperatures. And indeed, such a method is provided by one of the recent most active fields in material sciences, i.e., the sol/gel route to glass formation [8,17].

The main advantage of this technique for material scientists has been the ability to form composite oxide glasses (by choosing the proper monomers), which are impossible to obtain by the melting technique because of phase separation problems [9]. By this technique, glasses are prepared in two stages: First, a monomer, $Si(OCH_3)_4$ in our case, is polymerized at room temperature to form a solid dry porous gel glass [7,8,17]; and second, the pores are closed by heating to several hundreds of degrees. Since the pore-collapsing stage is still at elevated temperatures, the main question is, does the porous gel-glass show the desired properties of matrix isolation ? The answer is positive, and is detailed below. Trapping of the molecule is then performed by adding it to the starting polymerizing solution. Despite the simplicity of this idea, we are unaware of previous reports in which photoactive glasses have been made by this technique. Recently, Tani, Makishima et al., reported a photochemical hole-burning study of 1,4-dihydroxyanthraquinone trapped in silica by the sol/gel method [10].

The main properties of the silica glass are the following:
(a) The glass is transparent to U.V., down to 250 nm (Fig. 1 in ref. [3]).

In contrast, the plastics used for these purposes absorb in the near U.V. Porosity is no obstacle for transparency because pore size distribution is in the few nanometer range [11], much below U.V.-Visible wave-lengths.
(b) The molecules are trapped in silica cages. They are not adsorbed on the walls of the open pores. This has been shown by leaching experiments [3,5].
(c) Molecular isolation is very efficient [3,5-7]. High concentrations may be reached without aggregation.
(d) The silica network, $(SiO_2)_n$, and consequently also the cages, are much more rigid than plastic matrices.
(e) Since intra-matrix diffusion virtually does not exist, initial im-purities and photo-decomposition products are isolated and do not interfere.
(f) The silica glass is photochemically and thermally inert. This is a severe problem with plastic carriers [12].
(g) The doped glasses can be prepared in any desired geometry and shape, as detailed below (Fig. 1). Preparation of fluorescent doped thin-films has been described in detail elsewhere [5].

There are also a number of disadvantages: First, we are not certain yet whether even after thorough drying, a few water or methanol molecules do not still remain in the cages. Second, the glasses are brittle. Whereas this is no problem for laboratory purposes, any future commercialization idea will have to take care first of that glass characteristic. Third, for some applications, the rigidity of the cage may be, in fact, too high, preventing even intramolecular isomerizations [13].

A very wide variety of photoactive organic molecules have been trap-ped in silica and silica-titania glasses by the sol/gel technique [3,5-7, 13,14]. The method is, of course, not limited to organic molecules: in-organic salts and complexes can be trapped as well [4, 14].

The doped glasses can be prepared under acidic [4,5,10,14], basic [3,5,14] and neutral conditions [14]. A typical procedure for neutral conditions is given below.

2. Preparation of Blocks in which Rhodamine 6G (R6G) and B (RB) are Trapped

Trapping of R6G in silica glass has been described in detail previous-ly [3]. It has been found that fluorescence photostability is greatly improved, and that high concentration of the dye (over 10^{-3} M) can be reached without fluorescence quenching aggregation.

Since in most dye-lasers, solutions of the dye are pumped through the electro-optical set-up, it seemed to us desirable to explore the possibility of replacing the solutions with glass blocks. In order to do that, one has first of all to develop the skill of obtaining these glasses in desirable geometries. Luckily, the shrinkage of the gels during the desiccation stages, follows exactly the geometry of the container. Fig. 1 shows a few examples of rectangular blocks, cylindrical blocks and disks, in which either R6G or RB were trapped. The photophysical proper-ties of these blocks will be reported elsewhere. The reactangular blocks are prepared as follows:

Experimental Procedure

Into a rectangular 1x1x3 cm plastic cell was placed a solution of 2.08 ml $Si(OCH_3)_4$, 1.0 ml aqueous solution of R6G or RB in the desired concentration and 2.5 ml CH_3OH. The cell is closed with Al foil in which a hole was perforated with a needle. The cell is left untouched for a

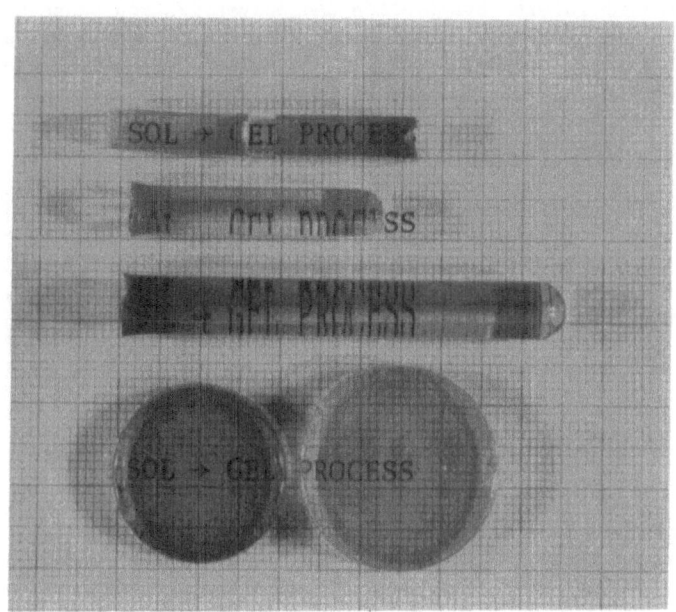

Fig. 1. Rectanuglar blocks (top), cylindrical rods (middle) and disks
(bottom) of rhodamine 6G (yellow) and rhodamine B (purple)
trapped in silica glass ($\sim 10^{-3}$ M).

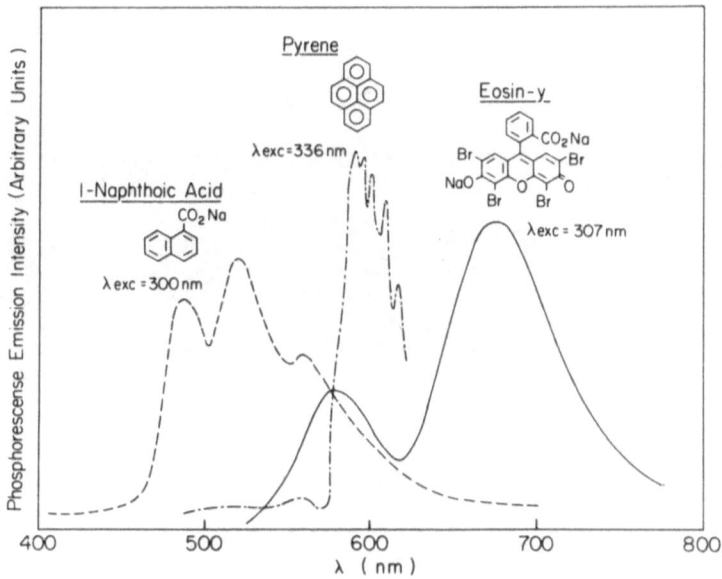

Fig. 2. Room temperature phosphorescence of molecules trapped in silica
glass ($\sim 10^{-3}$ M). --- 1-Naphthoic acid sodium salt; $-\cdot-\cdot-$ Pyrene
in the presence of 1M NaI; —— Eosin-y.

week at room temperature. Evaporation and polymerization cause shrinkage of the gel to 1/8 of the original volume. See Fig. 1 for the results. Other geometries are obtained (Fig. 1), by using other cell geometries. Shrinkage and desiccation time may vary.

3. Room Temperature Phosphorescence (RTP) of Trapped Molecules

Phosphorescence is observed under protective environments only (low temperatures or matrix isolation) [15]. We report now that the efficient caging described above, results in the ability to obtain RTP. Three examples are shown in Fig. 2: Py in the presence of a heavy atom, eosin-y and the Na salt of 1-naphthoic acid. It should be noted that Py phosphorescence on silica surfaces was observed only upon addition of ethanol and at 60°K [15]. Phosphorescence intensity and life-time increase with gel shrinkage and dryness [13].

The preparation of photochromic glasses has been described recently [6].

Acknowledgements: Supported by the NCRD, Israel, and by the KFA, Julich, West Germany. Assisted by the F. Haber Research Center for Molecular Dynamics, Jerusalem. We thank V.R. Kaufman for helpful discussions.

References

1. I.R. Dunkin, Chem. Soc. Rev., 1 (1974).
2. R. Reisfeld and C.K. Jørgensen, Structure Bonding, 49:1 (1982).
3. D. Avnir, D. Levy and R. Reisfeld, J.Phys.Chem., 88:5956 (1984).
4. D. Levy, R. Reisfeld and D. Avnir, Chem.Phys.Lett. 109:593 (1984).
5. D. Avnir, V.R. Kaufman and R. Reisfeld, J.Non-Cryst.Solids, 74:395 (1985).
6. V.R. Kaufman, D. Levy and D. Avnir, Proc. 3rd Int. Workshop on "Glasses and Glass Ceramics from Gels", Montpellier, Sept. 1985; J. Non-Cryst. Solids, in press, 1986.
7. V.R. Kaufman and D. Avnir, this volume.
8. S.Sakka and K. Kamiya, J. Non-Cryst.Solids, 42:403 (1980).
9. Vycor is an example of a phase-separated glass. See, D. Rojanski et al., this volume.
10. T. Tani, H. Namikawa, K. Arai and A. Makishima, J.Appl.Phys. in press; T. Tani, A. Makishima, A. Itani and U.Itoh, presented in this Symposium.
11. M. Yamane, S. Aso, O. Okano and T. Sakaiano, J.Mat.Sci., 14:607 (1979).
12. "Photochemistry of Dyed and Pigmented Polymers", N.S. Allen and J.F. McKellar, Eds., Appl. Sci. Publishers, London 1980.
13. D. Levy and D. Avnir, in preparation.
14. D. Levy, M.Sc. Thesis, The Hebrew University, 1984; D. Avnir, R. Reisfeld and D. Levy, Israel Patent Application 69794, Sept. 23rd, 1983.
15. R.T. Parker, R.S. Freedlander and R.B. Deulap, Anal.Chim.Acta 119:189 (1980); 120:1 (1980).
16. P. de Mayo, L.V. Natarajan and W.R. Ware, ACS Symp. Ser. 278:1 (1985).
17. H. Dislich, J. Non-Cryst. Solids, 57:371 (1983); C.J. Brinker and G.W. Scherer, ibid., 70:301 (1985).

OPTICAL MODULATION SPECTROSCOPY OF NATIVE DEFECTS IN AMORPHOUS

SEMICONDUCTORS AND CONDUCTING POLYMERS

Zeev Vardeny

Department of Physics and Solid State Institute
Technion - Israel Institute of Technology
Haifa 32 000, Israel

ABSTRACT

Basic characteristics of the photomodulation spectroscopy for dangling bond defects in semiconductors are discussed. The method is applied to studying the energy levels and the effective correlation energy for native defects in polyacetylene, a-Si:H and chalcogenide glasses.

INTRODUCTION

Illumination of a semiconductor changes the distribution of electrons, which leads to a variety of effects underlying the optoelectronic applications of these materials. These effects can also be used as tools for studying the electron states and the carriers dynamics. In this paper we will discuss the use of optical modulation (OM) spectroscopy for studying energy levels of native defects in two prototype amorphous semiconductors: a-Si:H and a-As$_2$S$_3$, and polyacetylene-$(CH)_x$ which is the simplest polymer among the conducting polymers.

Photomodulation uses two beams,[1] one beam for producing the excited states (pump) and another beam for measuring the changes ΔT in the transmission T (probe). The experimental set up for measuring ΔT over the range from 0.1 to 2.5 eV is shown in Fig. 1. The pump beam is an Ar$^+$ laser (sometimes used to pump a dye laser for excitation spectra measurements), chopped at about 150 Hz. In a typical experiment the absorbed flux is about 2 10^{18} photons/cm^{2}sec at $\hbar\omega$ = 2.41 eV. The transmission T and its photo-induced changes ΔT were measured using a tungsten lamp or a Perkin-Elmer Opperman source as probe beams. An ir monochromator and several semi-conductor detectors were used to span the whole energy range with optimal sensitivity. The sample temperature can be changed from 10 to 300K. At low temperatures, photoluminescence (PL) signal is inevitably collected with the ΔT signal; we measured the PL separately by blocking the probe beam and in the correction of the data we assumed simple superposition of ΔT and PL.[1] The data are plotted as $(-\Delta T/T) = d\Delta\alpha$ where $\Delta\alpha$ is the change in the absorption coefficient α and d is the sample thickness (assuming $d > 1/\alpha$).

ΔT are of two kinds,[2] photoinduced absorption (PA) and photoinduced bleaching (PB). PA is associated with transitions from initial states

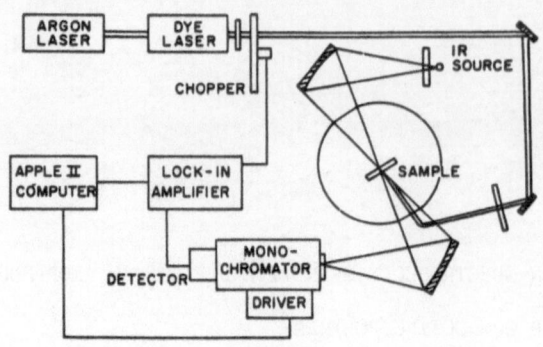

Fig. 1. The experimental set-up for steady
state photomodulation.

when their occupation by electrons is enhanced by the illumination, or alternatively, with transitions whose final states are emptied by the illumination. PB is associated with initial states occupied in the dark that the illumination empties, or with final states filled by the illumination. We will see that because of this variety of photoinduced changes, the OM spectra often contain more spectroscopic information than the fundamental optical spectra (although, for the interpretation of the OM spectra knowledge of the fundamental spectra seems necessary).

Since the illumination does not change the total density of electrons, it follows from the sum rule for α^3 that

$$\int \Delta\alpha(\omega)\,d\omega = 0 \qquad\qquad (1)$$

This equation shows that if the OM spectrum contains spectral intervals in which $\Delta\alpha > 0$ (PA), it must contain intervals in which $\Delta\alpha < 0$ (PB). Both PA and PB are the fundamental features of any OM spectrum.[2]

OM SPECTRA OF DANGLING BONDS

We will consider a dangling bond that has three states with energies in the forbidden gap[4]: D^+ (unoccupied), D^0 (occupied with one electron) and D^- (occupied with 2 electrons). The energy diagram of these states is shown in Fig. 2. If the lattice relaxation is negligibly small, the energies of the D^0 and the D^+ states are identical and the energies of D^- and D^0 differ by the bare correlation energy U.[4] The lattice relaxation changes the energies of D^- and D^+ states. In Fig. 2 it is assumed that D^- and D^+ relax by the same energy $1/2\ E_r$. As apparent from Fig. 2, the optical transitions t_1 and t_4 differ from the transitions involving bare states by E_r.[2]

While U is always positive, the "effective correlation energy" defined[4] by the reaction $2(D^0)_r + U_{ef} = (D^+)_r + (D^-)_r$, where $U_{ef} = U - E_r$, can have both signs. Figure 2 corresponds to $U_{ef} > 0$. In this case the ground state is a D^0 state which produces ESR, and the spectrum in the dark, $\alpha_0(\omega)$, contains transitions t_2 and t_3. Upon illumination, the D^0 states capture electrons and holes and they are transformed into ESR inactive D^+ and D^- states. Transitions t_2 and t_3 are bleached (PB) and transitions t_1 and t_4 are generated (PA). In chalcogenide glasses U_{ef} is often negative[5-7] and the lattice relaxation effects are so strong that the energy of $(D^+)_r$ is higher than that of $(D^-)_r$. In this case the charged states form the ground

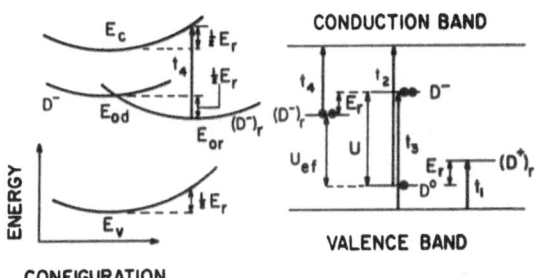

Fig. 2. Energy states of a dangling bond
        defect showing (a) configurational
        diagram and (b) optical transitions.
        E_{od} and E_{or} are the unrelaxed and
        the relaxed energies of the defect.

state which has no ESR response.[5,6] The OM spectrum contains transitions
t_1 and t_4 as PB bands and transitions t_2 and t_3 as PA bands.

For both cases, transitions t_1, t_2, t_3 and t_4 obey the relations[2]

$$t_3 - t_1 = t_2 - t_4 = U - E_r = U_{ef} \qquad (2)$$

$$t_1 + t_2 = t_3 + t_4 = E_g + E_r \qquad (3)$$

If the gap E_g is known, all the important parameters U_{ef}, E_r and U can
be determined from the OM spectrum, if at least three of the above tran
sitions are observed. If the energy levels are symmetric with respect to
the middle of the gap, knowledge of only two of the above transitions is
sufficient.

In order to determine the value of the effective correlation energy we
often use the sign of the correlation energy suggested by other experiments: ESR which measures the ground state spin properties and light induced ESR (LESR) which measures the excited state spin properties. If
$U_{ef} > 0$ there should be an ESR signal, while LESR shows bleaching of this
ESR signal, since the paramagnetic D^0 states are transformed into the diamagnetic $(D^+)_r$ states. For $U_{ef} < 0$ there is no ESR signal, however the
LESR signal is positive, since the $(D^+)_r$ states are transformed into D^0
under the illumination. In some special cases, including a-Si:H and $(CH)_x$,
it is not necessary to obtain U_{ef} sign from other experiments, since the
OM spectrum is consistent with only one sign of U_{ef}. Two additional
equations can be written[8] for the transitions t_n (Fig. 2)

$$t_1 + t_4 = E_g + E_r - U_{ef} \qquad (4)$$

$$t_2 + t_3 = E_g + U \qquad (5)$$

If $U_{ef} < 0$ t_2 and t_3 are the PA bands and the sum of the two PA
thresholds is larger than the gap (Eq. (5)). For $U_{ef} > 0$ t_1 and t_4 are the
PA bands and from Eq. (4) the sum of the two PA thresholds (Eq. (4)) can be

either larger or smaller than the gap. Thus, if experimentally observed PA bands have a sum less than the gap, the defect must have a positive U_{ef}.[8] This criterion is even simpler if charge conjugation symmetry exists. In this case there is a single PA band; if the PA threshold is smaller than $1/2 \ E_g$ the defect has a positive U_{ef}.

It is not always possible to determine the sign of U_{ef} from the OM spectrum. In particular, for every negative U_{ef} interpretation there is an equally consistent positive U_{ef} interpretation.[8] Denoting by a -(+) superscript a consistent interpretation of an OM spectrum with negative (positive) U_{ef}, they are related as follows:[8]

$$U_{ef}^+ = -U_{ef}^-; \quad E_r^+ = E_r^-; \quad U^+ = -U^- + 2E_r.$$

The strong condition that the relaxed D^- and D^+ states lose the same amount of relaxation energy (as assumed in Fig. 2) is unjustified if the two charged states have different coordination, as in the case of chalcogenide glasses.[7] In this case, all four transitions t_1 to t_4 are needed to obtain U_{ef}. Assuming that the curvature of the three states $D^{\pm}$ and D^0 are approximately equal, U_{ef} is given by the relation

$$U_{ef} = \frac{t_2 + t_3 - (t_1 + t_4)}{2} \tag{6}$$

The OM spectrum is the convolution of the defect density of states (DOS) with the continuum band DOS.[2] In trans-$(CH)_x$ the one-dimensional structure[9] leads to a $(E)^{-\frac{1}{2}}$, singularity for the DOS at the band edge. In this case, the OM spectrum is approximately proportional to the somewhat broadened DOS of the defect. In three dimensions, the spectrum lacks the sharp features. If the DOS in the continuum is a square-root function $(E)^{\frac{1}{2}}$, it can be roughly approximated by a step function. Then the OM spectrum is approximately proportional to the integral of the defect DOS, and therefore at the energy difference between the defect DOS maximum and the band edge, a *change in the slope* of the OM spectrum occurs.

We will now apply these ideas to measure energy levels and U_{ef} for the dangling bond native defects in three important semiconductors: trans-$(CH)_x$, the simplest quasi-one-dimension (1D) semiconductor (which is a conjugated polymer), in a-Si:H - the technologically important tetrahedrally bonded amorphous semiconductor and two of the chalcogenide glasses in which $U_{ef} < 0$ was predicted.

OM SPECTRUM OF SOLITONS IN TRANS-$(CH)_x$

Whithin the widely used model Hamiltonian[10] which treats trans-polyacetylene in terms of a coupled electron-lattice system, solitons are a primary excitation of the polymer chain. The soliton defect in $(CH)_x$ is a 1D domain wall[10,11] which separates the two degenerate ground state structures of the trans isomer. As a result of their translational invariance which leads to a small kinetic mass[10], solitons are thought to play the role of energy and charge carrying excitations. Solitons possess an unusual spin and charge assignments: the charged solitons $S^-(S^+)$ localize a pair of electrons (or holes) and hence have spin 0, while neutral solitons S^0 have spin 1/2. These defects exist in films of trans-$(CH)_x$ with concentration of about 10^{19} cm^{-3} as a result of isomerization from cis-$(CH)_x$.[9]

In a series of beautiful experiments, the transient sub-band gap optical absorption due to photoexcited states has been extensively characterized.[12-15] At low temperatures two distinct PA bands were observed with peaks at 0.45 eV and 1.35 eV. The PA band at 1.35 eV is due to an overall

neutral state which is intrinsically formed; it decays within a few microseconds and sdisappears completely from the steady state OM spectrum for
temperature higher than 200K. This state has been identified[12,15] as a
photoexcited soliton-antisoliton pair, or a breather mode.[16] The candidate
for the photoproduction of S^+ is the PA band at 0.45 eV. However, it was
demonstrated[14] that this band is extrinsically photoproduced and its
strength critically depends on the density of S^0 in the sample. Therefore
it was argued[14] that S^+ can be photogenerated mainly via the reaction
$2S^0 \rightarrow S^+ + S^-$. This process was identified by LESR[17] which shows photo-
quenching of the ESR signal associated with S^0.

The full OM spectrum of the soliton defects in trans-$(CH)_x$ film
($d \simeq 1000$ Å) kept at 210K, is shown in Fig. 3. Due to the 1D character of
$(CH)_x$ the bands are sharply defined. Two well defined bands, a PA and a PB
bands, are seen and the sum-rule for $\Delta\alpha$ (Eq.(1)) is approximately obeyed.
This also shows that the two bands share a common origin. The additional
modulation around the PB peak is probably caused by vibronic side-bands and
we identify the shoulder at 1.45 eV as the zero-phonon transition. Associated with the PA band, which peaks at 0.5 eV at this temperature, is a
narrow photoinduced ir-active vibration at 0.17 eV which shows[13] that the
PA band is due to photoinduced charged defects $(S^\mp)$. A sharp feature in
the PB band that could be associated with the bleaching of an ir-active
vibration does not exist and therefore the PB band is due to the bleaching
of neutral defects (S^0) transitions. Charge conjugation symmetry[18] is
evident from the OM spectrum of Fig. 3, since we do not observe separate
PA bands arising from S^+ and S^-. From the criterion which sets the sign of
U_{ef}[8] (Eq. (4)), we conclude that $U_{ef} > 0$ for the soliton defect, since
$t_1 + t_4 = 2 \times 0.5 = 1$ eV $< E_g$ ($\simeq 1.7$ eV).

The optical transitions t_1 to t_4 associated with the soliton defect
can be readily identified from the OM spectrum of Fig. 3. For the charged
soliton (PA) $t_1 = t_4 = 0,5$ eV, while for the neutral soliton (PB)
$t_2 = t_3 = 1.45$ eV. These values show that the soliton defect in trans-$(CH)_x$
cannot be described in terms of electron-lattice interaction alone, since
in this approximation[10] the energies of all four transitions t_1 to t_4 are
equal to $1/2 \ E_g$ (approximately 0.9 eV). Using the 1D energy gap of 1.7 eV

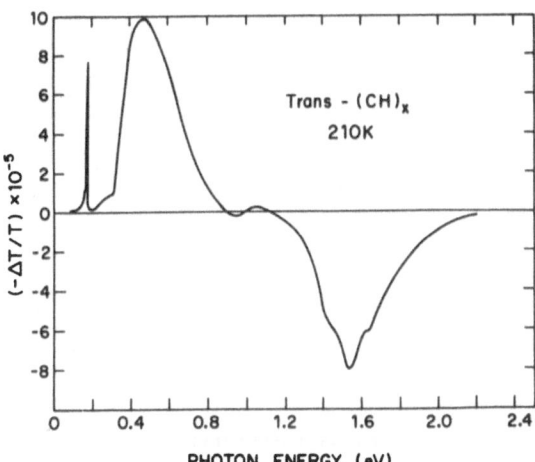

Fig. 3. OM spectrum of the soliton defects
in trans-$(CH)_x$ at 210K.

and Eqs. (2) and (3), we calculate for the soliton defect $U_{ef} = t_3 - t_1 = 0.95$ eV, $E_r = E_g - (t_1 + t_2) = 0.25$ eV, and $U = U_{ef} + E_r = 1.2$ eV. The large magnitude of U compared with E_g indicates that the e-e interaction must be included in any realistic theory of excited states in $(CH)_x$.

The OM spectrum of the soliton defect shown in Fig. 3 contains another interesting feature which was not discussed before. Even that one PA and one PB bands are photoproduced and thus charge conjugation symmetry holds,[18] these bands are asymmetrically displaced from the midgap ($1/2E_g \simeq 0.85$ eV): the PA band is displaced less from the midgap than the PB band. In any perturbation theory used to describe the e-e interaction,[18] a symmetrical displacement from midgap is predicted for $S^\mp$ and S^0 transitions. The apparent puzzle can be readily solved if we allow for lattice relaxation. Due to the Coulomb repulsion between the two like charges the $S^\mp$ length is higher than that of S^0. Assuming that the e-e interaction depends on the soliton length, this interaction is weaker for the S^+ compared to S^0 and therefore $S^\mp$ transition is displaced less from the midgap than the S^0 transition.

OM SPECTRUM OF DANGLING-BONDS IN a-Si:H

The most important defect in undoped a-Si:H is the three-fold coordinated Si atom, or dangling bond (DB) defect T_3. The most common ESR signal seen in undoped a-Si:H, with a g value of 2.0055, is believed to be the signature of the neutral T_3^0 defect;[20] thus $U_{ef} > 0$ for the DB defect in a -Si. The charged states, T_3^{-3} or T_3^+, are diamagnetic and hence do not contribute to the LESR signal. The occurrence of light-induced reaction $2T_3^0 \rightarrow T_3^- + T_3^+$ was identified[21] by LESR in a-Si:H with high density of DB.

Our a-Si:H film was a high quality glow discharge grown material which was electron irradiated so that the density of T_3^0 defects measured by ESR was 2×10^{18} cm^{-3}. Since the DB density is so high, we expect that transitions involving these defects will dominate the OM spectrum while the contribution of the band-tails, often observed in a-Si:H,[1] is negligibly small.

The OM spectrum of the electron irradiated a-Si:H at 10K is shown in Fig. 4. It consists of two PA bands followed by a reduction of absorption. These bands are not as sharply defined as in $(CH)_x$ (Fig. 3), because in 3D

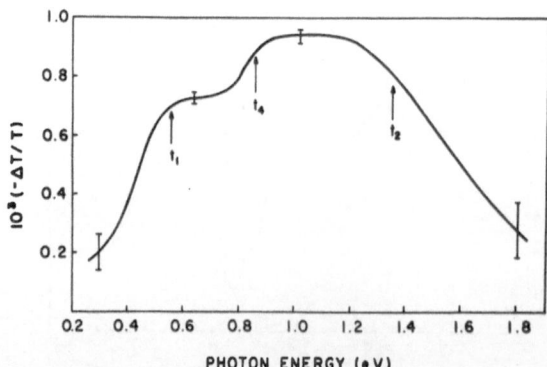

Fig. 4. OM spectrum of the dangling bond defects in electron irradiated a-Si:H at 10K.

202

the DOS of the continuum does not have a singularity at the band edge, as in 1D. Nevertheless, as explained in §2, we can estimate the energy levels from the change of the PA band slope. For PA, the changes of slope occur at 0.55 and at 0.85 eV. The onset of the bleaching is at 1.2 eV; if we assume the same band width as that of the PA bands, we estimate that the PB energy level is at 1.35 eV. The three bands in the OM spectrum share a common origin as confirmed by their dependence on the pump intensity I_L; all of them follow $I_L^{\frac{1}{2}}$ dependence.

We identify the energies 0.55, 0.85 and 1.35 eV as transitions t_1, t_4 and t_2, respectively, involving the DB defects. Transition t_2 seen in the absorption spectrum of a-Si:H[22] is bleached upon illumination and two PA bands appear, as T_3^0 become charged and T_3^+ and T_3^- are formed. These results show that the T_3^+ level (t_1) peaks at 0.55 eV above the valence band and the T_3^- level (t_4) peaks at 0.85 eV below the conduction band, in agreement with the DLTS data[23] and the optical absorption spectrum of doped materials.[24] The asymmetry of T_3^+ and the T_3^- energy levels with respect to the middle of the gap is evident from the OM data (Fig. 4). This indicates that the conduction and valence band edges are not formed from the same interaction (as is the case for $(CH)_x$) and hence charge conjugation symmetry does not exist for a-Si:H. Using Eq. (2) we obtain for the DB defect $U_{ef} = t_2 - t_4 = 0.5$ eV which is close to the values of U_{ef} determined from the absorption spectrum[24] ($U_{ef} = 0.35$ eV) and ESR[25] ($U_{ef} = 0.4$ eV). Since $E_g = 1.75$ eV at low temperature, we conclude from Eq. (3) that $E_r = t_1 + t_2 - E_g \simeq 0.1$ eV. This shows that the relaxation around an isolated DB defect which was created by electron irradiation is very small, but it does not contradict the idea that the relaxation around a $T_3^+T_3^-$ pair, or a T_3^+ defect associated with an ionized dopant atom, is indeed large.[4]

A second consistent interpretation of the OM spectrum shown in Fig. 4 was recently suggested.[8] It is based on a different identification of the PB onset at 1.35 eV. Identifying it as the transition t_3 instead of t_2 (as originally suggested[2]), the values for U and U_{ef} can be again calculated using Eqs. (2) and (3), and they are U = 1.3 eV and $U_{ef} = 0.8$ eV. These values are too large and they are not in agreement with other experiments.[23-25] In addition, the fourth transition (t_2) which is consistent with this set of values is predicted[8] to be at 1.65 eV, but none is observed (Fig. 4).

OM SPECTRA OF IVAP IN CHALCOGENIDE GLASSES

The idea that $U_{ef} < 0$ for the defects in the chalcogenide glasses was introduced by Anderson[5] to explain the pinning of the Fermi level and the absence of spin paramagnetism. Specific models[6,7] such as the valence alternation pairs (VAP) were developed to explain why $U_{ef} < 0$. The defect ground state is charged D^- or D^+ and therefore it is diamagnetic. Upon illumination the reactions $D^- + h \rightarrow D^0$ and $D^+ + e \rightarrow D^0$ may take place inducing paramagnetism and mid-gap absorption. These ideas were experimentally confirmed by Bishop, Strom and Taylor[26] who observed LESR signals with spin densities up to 5×10^{17} cm^{-3}, photoinduced optical absorption in the gap and PL fatigue.

Later, it was shown[27] that spin densities of 10^{20} cm^{-3} can be photoinduced in a-As$_2$S$_3$. These measurements suggested that there are two sets of defects, the isolated defects ($D^\mp$) and the photoinduced defect pairs D^+D^- (intimately related VAP[7], called IVAP). These pairs are amphoteric, since they can capture either an electron ($D^+D^- + e \rightarrow D^0D^-$) or a hole ($D^+D^- + h \rightarrow D^+D^0$), inducing paramagnetism and mid-gap absorption. While the excited isolated defect is metastable,[26] a subset of the excited D^+D^- pairs can decay to the ground state much more quickly, as observed in the optically detected magnetic resonance.[28] Since the recombination is faster than our chopping frequency (150 Hz), the OM technique can be applied to determine the defect energy levels.

There are four optically induced transitions related to each IVAP (Fig. 2); E_r is very large so that $(D^+)_r$ is close to the CB edge and $(D^-)_r$ is close to the VB edge. The OM spectrum gives the spectrum associated with PA from D^0D^- into the CB (t_2) and from the VB into D^+D^0 (t_3), and PB associated with the ground state defect D^+D^- (t_1 and t_4).

The OM spectrum measured at 10K on a sample of a-As_2S_3 which was 100 µm thick, is shown in Fig. 5. Only one OM band is observed with onsets at 1.3 eV for PA and 2.3 eV for PB. This indicates either that one charged state ($(D^+)_r$ or $(D^-)_r$) is optically active, or that charge conjugation symmetry exists in a-As_2S_3. We adopt the latter explanation in agreement with the suggestion of Mott et al;[6] and therefore assign the PA onset as t_2 = t_3 = 1.3 eV and the PB onset as t_1 = t_4 = 2.3 eV, within 0.1 eV. Using E_g = 2.5 eV at 10K and Eqs. (2) and (3), we calculate E_r = 1.1 eV, U = 0.1 eV and U_{ef} = -1.0 eV. This directly confirms that U_{ef} < 0 for the IVAP defect in a-As_2S_3. The second interpretation of the OM identifying the 1.3 eV onset as t_2 or t_3, result in E_r = 1.1 eV; U_{ef} = 1.0 eV (>0), but U = 2.1 eV which is much too large to be acceptable given the high dielectric constant of a-As_2S_3. Our results also give the energy of the shallow donor and acceptor levels from the nearest band edges (0.2 eV) which is also in agreement with the suggestion of Mott and Street.[6]

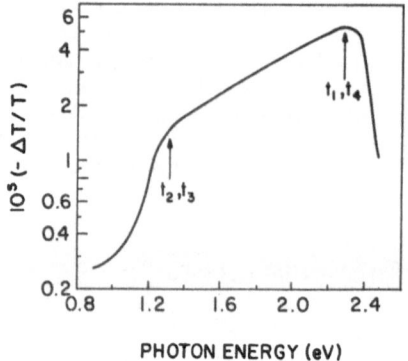

Fig. 5. OM spectrum of the D^+D^- pair defects in a-As_2S_3 at 10K.

We did the same analysis for a-As_2Se_3 using the transient OM spectrum at 20K which was published by Orenstein and Kastner.[29] From the onset of PA and PB we find t_2 = t_3 = 1 eV, t_1 = t_4 = 1.7 eV and since E_g = 1.9 eV at 20K, we calculate E_r = 0.8 eV, U = 0.1 eV and U_{ef} = -0.7 eV, in agreement with U_{ef} recently determined[30] from transient PC of a-As_2Se_3. Our results show that although U (0.1 eV) and the shallow donor and acceptor energies (0.2eV) are equal in both a-As_2Se_3 and a-As_2S_3, U_{ef} is quite different, because the lattice relaxations are different.

ACKNOWLEDGEMENTS

I thank the BSF Foundation, Jerusalem, Israel, for the financial support which made my stay at Brown University possible.

I thank J. Tauc for many helpful discussions and for the a-As$_2$S$_3$ sample. I also thank H. Dersch for the irradiated a-Si:H film and J. Tanaka - for the (CH)$_x$ sample. This work was partly supported by the Israel Academy for basic research.

REFERENCES

1. P.O'Connor and J. Tauc, Phys. Rev. B 25, 2748 (1982).
2. Z. Vardeny and J. Tauc, Phys. Rev. Lett. 54, 1844 (1985).
3. M. Altarelli, D.L. Dexter, H.M. Nussenzveig, D.Y. Smith, Phys. Rev. B 6, 4502, 1972.
4. D. Adler , in: "Semiconductors and Semimetals", ed. J.I. Pankove, New-York; Academic Press, Vol. 21A, p. 291 (1984).
5. P.W. Anderson, Phys. Rev. Lett. 34, 953 (1975).
6. R.A. Street and N.F. Mott, Phys. Rev. Lett. 35, 1293 (1975).
7. M. Kastner, D. Adler and H. Fritzsche, Phys. Rev. Lett. 37, 1504 (1976).
8. Y. Bar-Yam, J.D. Joannopoulos and D. Adler, Phys. Rev. Lett. 55, 138 (1985).
9. A.J. Heeger, Comments on Solid State Physics 10, 53 (1981).
10. W.P. Su, J.R. Schrieffer and A.J. Heeger, Phys. Rev. Lett. 42, 1698 (1979); Phys. Rev. B 22, 2099 (1980).
11. M.J. Rice, Phys. Lett. A 71, 152 (1979).
12. J. Orenstein and G.L. Baker, Phys. Rev. Lett. 49, 1043 (1982).
13. Z. Vardeny, J. Orenstein and G.L. Baker, Phys. Rev. Lett. 50, 2032 (1983).
14. J. Orenstein, Z. Vardeny, G.L. Baker, G. Eagle and S. Etemad, Phys. Rev., B 30, 786 (1984).
15. Z. Vardeny, Physica B 127, 338 (1984).
16. A.R. Bishop, D.K. Campbell, P.S. Lomdahl, B. Horovitz and S.R. Phillpot, Synthetic Metals 9, 223 (1984).
17. J. Orenstein, Proc. Int. Conf. on the Physics of Semiconductors, ed. J.D. Chadi and W.A. Harrison, Springer-Verlag, 1984, p. 1447.
18. S. Kivelson and W.K. Wu, Mol. Cryst. Liq. Cryst. 118, 9 (1985).
19. D.K. Campbell, T.A. DeGrand and S. Mazumdar, Phys. Rev. Lett. 52, 1717, (1984).
20. P.J. Caplan, E.M. Poindexter, B.E. Deal, R.R. Rozouk, J. Appl. Phys. 50, 5847 (1979).
21. A. Friederich and D. Kaplan, J. Non-Cryst. Solids 35/36, 657 (1980).
22. W.B. Jackson and N. Amer, Phys. Rev. B 25, 5559 (1982).
23. J.D. Cohen, J.P. Harbison and K.W. Wecht, Phys. Rev. Lett. 48, 109 (1982).
24. W.B. Jackson, Sol. Sta. Commun. 44, 477 (1982).
25. H. Derch, J. Stuke and J. Beichler, Phys. Stat. Solidi (b) 105, 265 (1981).
26. S.G. Bishop, U. Strom and P.C. Taylor, Phys. Rev. Lett. 34, 1346 (1975); Phys. Rev. B 15, 2278 (1977).
27. D.K. Biegelsen and R.A. Street, Phys. Rev. Lett. 44, 803 (1980).
28. S.P. Depinna and B.C. Cavanett, Phys. Rev. Lett. 48, 556 (1981).
29. J. Orenstein and M. Kastner, Phys. Rev. Lett. 46, 1421 (1981).
30. T. Thio, D. Monroe and M. Kastner, Phys. Rev. Lett. 52, 667 (1984).

MICROSTRIPLINE TRANSIENT PHOTOCURRENTS: A PROBE OF CARRIER THERMALIZATION AND BAND-TAIL STATE DENSITIES IN AMORPHOUS SEMICONDUCTORS

T.E. Orlowski and M. Abkowitz

Xerox Webster Research Center
800 Phillips Road, Webster, NY 14580

INTRODUCTION

Much progress has been made concerning the general features of carrier transport and thermalization mechanisms in amorphous semiconductors from time-resolved studies of photocurrents, photoluminescence and photoinduced absorption. A common feature of all of these measurements is the observation of anomolous dispersion in the carrier dynamics characterized by a power-law time decay which is generally thought to derive from an exponential distribution (energy) of band-tail states in which thermalization procedes by multiple trapping (MT). However, an exponential energy distribution of hopping sites will also give rise to a power-law decay. Therefore, a determination of the transport and thermalization mechanisms operative in amorphous semiconductors is a difficult problem compounded by uncertainty regarding the actual energy distribution of band-tail states **near the band edges** and whether or not any structure exists. The purpose of this paper is to present new transient photocurrent measurements in a-Se and a-Si:H with improved experimental time resolution which allows one to examine carrier thermalization and the nature of the distribution of states close to the band edges.

EXPERIMENTAL

Transient photocurrents (TP) were studied in thin (3000Å) films of a-Si:H (rf glow-discharge, 250 mtorr SiH_4, 250°C. quartz substrate) and a-Se (vacuum evaporation, 55°C. quartz substrate). To insure maximum frequency response, coplanar metal electrodes (~200 μm gap) of Cr-Al were deposited as a microstrip 50Ω transmission line and a Cr-Al ground plane was deposited on the underside of the substrate. The gap was uniformly illuminated with 15 psec pulses from a picosecond dye laser approximately 5 μsec following application of the bias field (10 μsec pulse) to one of the electrodes. In this manner, nonuniform field effects due to dielectric relaxation were avoided. Photocurrent transients were amplified and transmitted to a 1 GHZ waveform digitizer coupled to a laboratory computer. Photocurrents were found linear over the entire experimental range of applied fields (0.2↔4.0 V/μm). The peak photocurrent was found linear in excitation density , N_p, and the TP decay independent of intensity for $N_p < 3 \times 10^{15}$ absorbed photons/cm^3. For larger N_p, bimolecular recombination influenced the carrier population on the timescale of these measurements. The maximum photogenerated charge in these experiments was 4×10^{-12} C, low enough to avoid space-charge effects.

RESULTS AND DISCUSSION

Figure 1 shows a log-log plot of the initial decay (t < 100 nsec) of the electron photo-current, i(t), in a-Si:H at 300K and 170K averaged over 256 laser shots with laser excitation at 2.2 eV. Also shown are fits of a power law, $i(t) \propto t^{-(1-\alpha)}$ to the data. The power-law nature of the TP decay observed here is characteristic of anomalous dispersion in the charge transport process and was first sucessfully explained by the continuous-time-random-walk (CTRW) theory developed by Scher and Montroll[1]. The anomalous transport is described by a dispersion parameter, α, which determines the exponent in the waiting time distribution function, $\Psi(t) \propto t^{-(1-\alpha)}$. Here, $\Psi(t)$ represents the distribution of hopping or waiting times in the transport process which can arise from a distribution of hopping distances or trapping state energies. One way of obtaining information concerning the transport mechanism and the nature of the distribution (whether spatial, energetic or both) giving rise to the dispersion is to examine the temperature dependence of α. Measurements of $\alpha(T)$ in a-Si:H corresponding to the initial TP decay (t < 100 nsec) were performed over the range 150 to 320K and determined that $\alpha(T) = \alpha_0 + \beta T$ with $\alpha_0 = 0.44$ and $\beta = 0.001$ K^{-1}. Although the linear temperature dependence found here for α is consistent with thermalization involving MT in an exponential band-tail, the non-zero value of α at T=0 is not. Similar behavior is found for the T-dependence of the initial decay

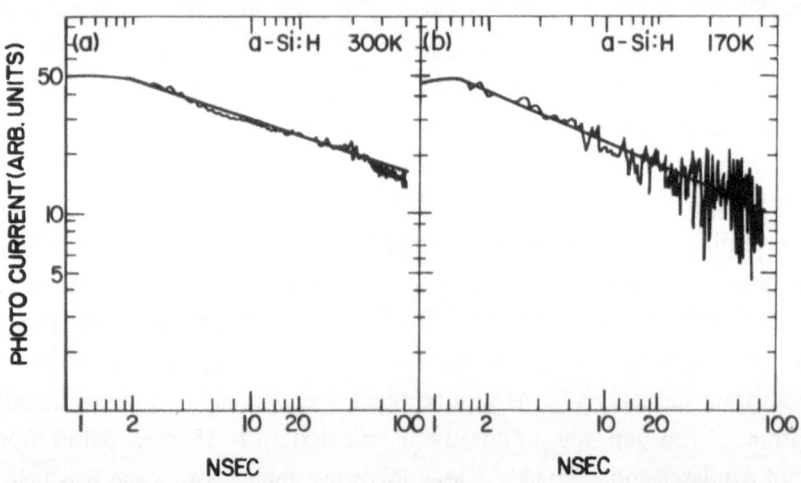

Fig. 1 Log-log plot of i(t) vs. t for a-Si:H at 300K and 170K. Also shown are fits to a power law, $i(t) \propto t^{-(1-\alpha)}$ where in a) $\alpha = .74$ and in b) $\alpha = .62$.

of photoluminescence[2] in a-Si:H. We find this temperature dependence of α incompatible with either a pure MT process or a simple hopping down process in an exponential band-tail as envisaged by Monroe[3]. It is proposed that the weak temperature dependence found here in a-Si:H for α at **short** times involving thermalization within **shallow** band-tail states requires 1) both MT and carrier hopping where hops upward of $\sim\kappa T$ are included in the thermalization process and/or 2) a different functional form of the density of band-tail states near the conduction band edge.

Figure 2 shows a log-log plot of the initial decay (t < 50 nsec) of the hole photocurrent, i(t), in a-Se at 295K and 270K averaged over 2000 laser shots with laser excitation at 2.6 eV. The initial decay is power-law like with some curvature at both temperatures. **Sharp changes in the slope are observed at $t_D = 11$ nsec at 295K and $t_D = 22$ nsec at 270K.** These results can be understood based upon recent Monte Carlo simulations[4] where transient photocurrents were calculated for an exponential band-tail with a discrete feature superimposed. Changes in the rate of photocurrent decay were correlated with the magnitude of the density of states due to the feature relative to the band-tail and the feature's width (energy) in units of κT. These results were later generalized[5] to obtain an analytical form for i(t) corresponding to a band-tail state density g(E) of the form

$$g(E) = g_V \exp\{ (E-E_V) / \kappa T_o \} + g_D(E)$$

where the first term refers to the exponential tail and $g_D(E)$ reflects the discrete feature state density. Three parameters are needed in this model to correlate g(E) and i(t). They are $\Gamma = N_D/N_V$, where N_D is the total number of discrete states $(N_D = \int g_D(E)dE)$ and $N_V = g_V \kappa T$; $\alpha = T/T_o$, the dispersion parameter, which characterizes the width of the band-tail and $t_D = \upsilon_o^{-1} \exp\{-E_D/\kappa T\}$, the time it takes the demarcation energy to reach the position (energy) of the discrete feature (denoted

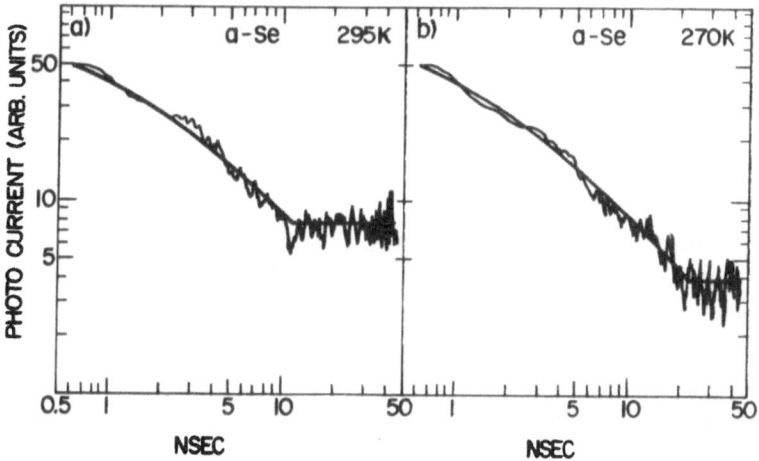

Fig. 2 Log-log plot of i(t) vs. t for a-Se at 295K and 270K. Also shown are fits to the model discussed in the text.

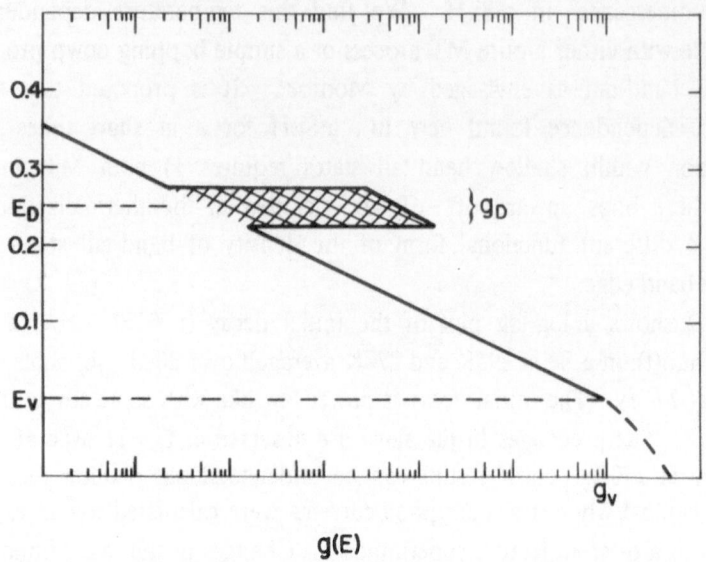

Fig. 3 Schematic diagram of the valence band-tail density of states, g(E), in a-Se obtained from fits of i(t) data in Fig. 2.

by E_D). Shown in Figure 2 are fits of this model to our a-Se data. At 295K we find $\alpha = 0.92$, $\Gamma = .016$ and $t_D = 11$ nsec while at 270K we find $\alpha = 0.87$, $\Gamma = .015$ and $t_D = 22$ nsec. These parameters imply that a-Se has an exponential valence band-tail characterized by $T_0 = 315K$ which has superimposed on it a strong discrete feature. Time-of-flight measurements[6] locate this feature between 0.25 and 0.30 eV above the valence band mobility edge. Using these values of E_D in the expression for t_D, one obtains values of ν_0, the characteristic escape frequency, in the range $(0.2 \leftrightarrow 1.5) \times 10^{13}$ sec^{-1}. Shown in Fig. 3 is a schematic diagram of g(E) in a-Se obtained from the fits to i(t) where the feature is set at 0.25 eV and the width is set arbitrarily at $2\kappa T$. These results demonstrate that structure in g(E) in the vicinity of the mobility edge can be observed through measurements of transient photocurrents, in conformity with predictions of recent models, if the time resolution is sufficiently high and that structure is prominent.

REFERENCES

1) H. Scher and E.W. Montroll, Phys. Rev. B **12** (1975) 2455.
2) T.E. Orlowski and H. Scher, Phys. Rev. Lett. **54** (1985) 220.
3) D. Monroe, Phys. Rev. Lett. **54** (1985) 146.
4) J.M. Marshall and R.A. Street, Solid State Commun. **50** (1984) 91.
5) M. Silver, E. Snow and D. Adler, Solid State Commun. **53** (1985) 637.
6) S.O. Kasap and C. Juhasz, J. Phys. D **18** (1985) 703; and references therein.

PICOSECOND DECAY OF PHOTOINDUCED ABSORPTION IN

HYDROGENATED AMORPHOUS SILICON

Dale M. Roberts, Joseph F. Palmer, and Terry L. Gustafson

Standard Oil Research & Development
4440 Warrensville Center Road
Warrensville Heights, Ohio 44128

INTRODUCTION

The study of the transient behavior of photogenerated carriers is helping to elucidate the mechanisms for carrier relaxation in amorphous silicon (a-Si) materials. The dynamics of hot carrier thermalization, localization, trapping, and recombination can all be probed using time resolved optical techniques. There are three primary methods for obtaining this information: photoconductivity (PC) probes carriers above the mobility edge;[1,2] photoluminescence (PL) measures radiative and non-radiative recombination rates;[3-12] and photoinduced absorption (PA) probes changes in the optical absorption due to the presence of excited carriers.[13-22]

PA dynamics of a-Si have been studied over many different time scales; most recently picosecond[13-17] and sub-picosecond[18] excitation have been used. PA decays on these time scales have been obtained for unhydrogenated a-Si,[18,19] and doped and undoped hydrogenated amorphous silicon (a-Si:H).[15,16,20,21] These data show evidence for a correlation between the dynamics and the defect density of the material. The data for the doped a-Si:H show evidence for photodegradation with exposure to the pump and probe beams.[21,22] The picosecond PA decays also show evidence for dispersive behavior.[23,24] Based on these observations, existing data have been fit to the multiple trapping model of Scher,[25] Tiedje and Rose,[26] and Orenstein and Kastner.[27] The PA decays are interpreted as the change in cross section for absorption as the excited carriers relax; the population of excited carriers remains the same.

The recent picosecond PA experiments have used the same wavelength for both the pump and the probe.[13-22] In this work we present picosecond PA results on a-Si:H that were obtained using independently tunable pump and probe lasers. We show that the observed PA decay, as monitored using both transmittance and reflectance, is different at different sample thicknesses. This distortion is caused by an etalon effect in the sample and will be present for all samples if the steady-state optical absorption spectrum shows interference fringes. We model the induced absorption using numerical differentiation of equations for the steady state transmittance and reflectance, T_0 and R_0. Although the dominant contribution to the induced transmittance (ΔT) is the change in absorption, $\Delta\alpha$, the change in the real part of the index of refraction, Δn, contributes in such a way as to distort the decay curve at different thicknesses. Likewise, the dominant contribution to the induced reflectance (ΔR) is Δn, but $\Delta\alpha$ distorts the ΔR decay curve at different sample thicknesses. We compare several methods for providing thickness independent decays for the charge carriers in a-Si:H. We interpret the observed decays using the model for dispersive transport. We also show that the PA decay depends on carrier density and source repetition rate.

EXPERIMENTAL METHODS

We have presented the details of the experimental apparatus elsewhere.[28-31] Briefly, a mode locked argon ion laser pumped two synchronously pumped cavity dumped dye lasers to produce independently tunable pump and probe pulses. The pump dye laser was R590, tunable from 1.94-2.18 eV; the probe dye laser was either R590 or DCM, providing tunability from 1.70-2.18 eV. The probe beam was directed along a variable delay and combined colinearly with the pump beam. The light transmitted through the sample was spectrally resolved with a prism; light reflected from the sample was spectrally resolved using a grating. For both transmittance and reflectance, the probe beam was detected with a photodiode. The cross correlation between the pump and probe pulses gave a time resolution of ~8 ps. The pulse repetition rate could be varied and was typically 500 kHz. The pulses were focused through a 100 mm focal length lens to a spot size of ~50 microns, with the pump being slightly larger in diameter at the focus than the probe.

We detected the increase in absorption using a time modulation technique in order to eliminate the thermal background that is present with mechanical chopping.[28,29] In this scheme the time position of the probe pulse is modulated with respect to the pump pulse. The two modulation states were probe before

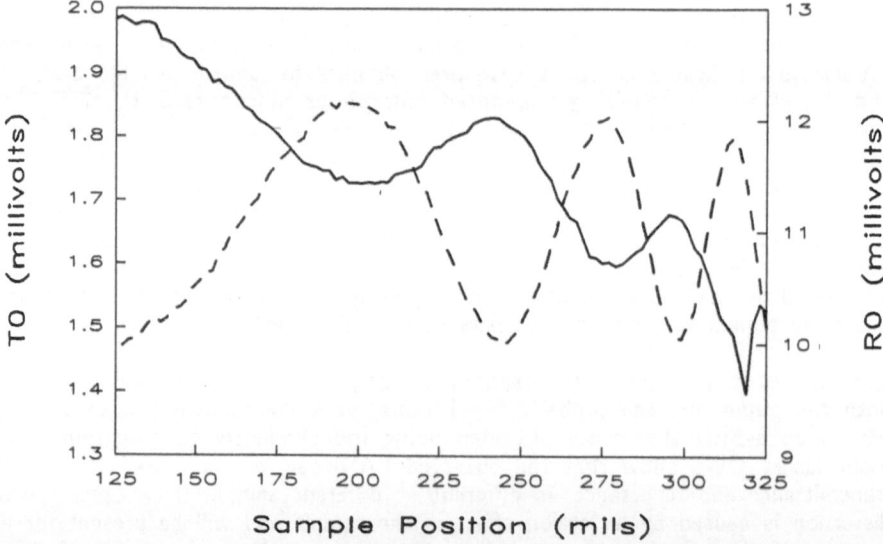

Figure 1. Steady state transmittance (solid) and reflectance (dash) of a-Si:H as a function of sample position: 2.11 eV pump; 2.02 eV probe; 2.4×10^{18} cm^{-3} carrier density; -100 psec delay.

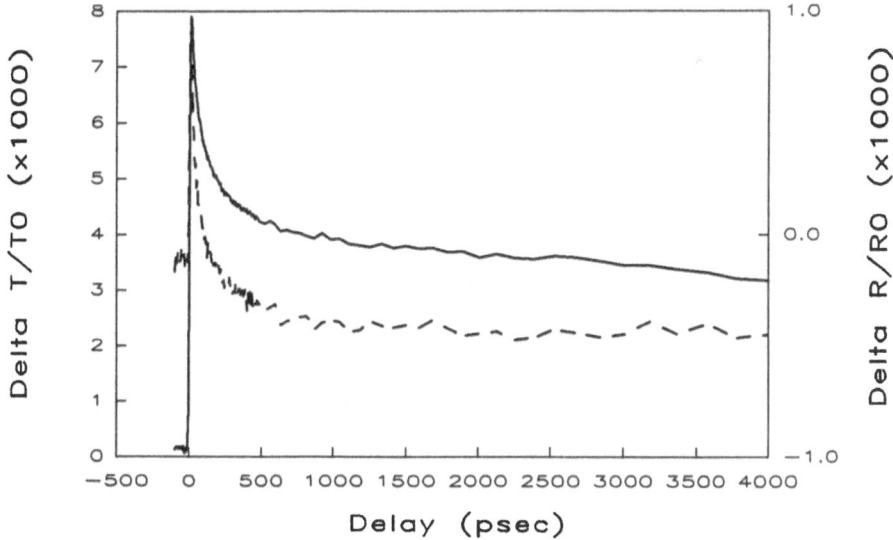

Figure 2. Decay of ΔT (solid) and ΔR (dash) at sample position 265. (See Fig. 1.) 2.11 eV pump; 2.02 eV probe; 2.4 x 10^{18} cm^{-3} carrier density.

pump (by 26 ns) and probe after pump (by the variable delay). In this way both the pump and probe pulses are present at the sample each pulse repetition period. This time modulation was accomplished by electronically switching the delay setting in the cavity dumper driver electronics. The picosecond PA signal was detected synchronously using the electronic switching rate as the reference. All decays were obtained at room temperature. Data collection times were typically 300 or 600 seconds.

The sample used in this work was deposited in a capacitively coupled rf glow discharge system. The sample was assessed as having a low defect density on the basis of the steady state photoluminescence intensity.[32]

RESULTS

In Figure 1 we show the transmittance and reflectance, T_0 and R_0, of the probe beam through the sample as a function of the sample position at a wavelength of 2.02 eV. In order to simulate steady state conditions for the transient measurement, these data were obtained with both the pump and probe incident on the sample with a relative delay of -100 psec; the pump generated a peak charge carrier density of 2.4 x 10^{18} cm^{-3}. It is clear that our sample was not uniform in thickness over the area we studied, but rather it increased in thickness monotonically. The intensity of the transmitted beam decreases as the sample position increases owing to the increasing thickness. And both T_0 and R_0 show interference fringes, with successive peaks corresponding to a sample thickness change of $\lambda/2$. Note that the peaks are narrower as position increases, indicating that the sample is increasing in thickness super-linearly with position. For every sample we have studied we have observed interference fringes caused by thickness variations.

In Figures 2 and 3 we show the decay of ΔT and ΔR at two sample positions at a charge carrier density of 2.4 x 10^{18} cm^{-3}. The pump and probe

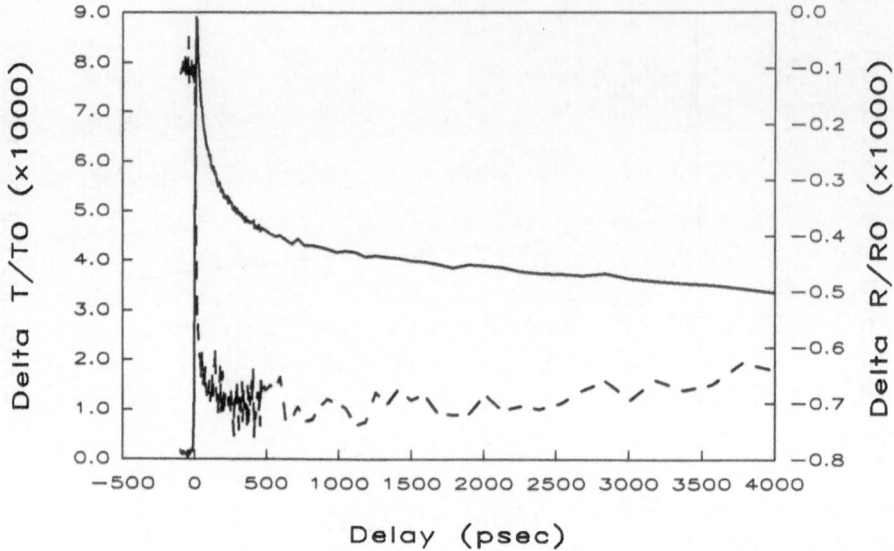

Figure 3. Decay of ΔT (solid) and ΔR (dash) at sample position 300. (See Fig. 1.) 2.11 eV pump; 2.02 eV probe; 2.4×10^{18} cm^{-3} carrier density.

wavelengths were 2.11 and 2.02 eV, respectively. The shape of the ΔT decays appear similar, but the ΔR decays are quite different in both shape and magnitude. (We note that *increasing* ΔT and ΔR correspond to *increasing* absorption.) The magnitude of $\Delta R/R_0$ is approximately one order of magnitude less than that of $\Delta T/T_0$. In addition, ΔR exhibits both increasing (absorption) and decreasing (bleaching) effects, even at the same sample position (see Fig. 2). In Figure 4 we compare the normalized decays of ΔT at four sample postions. Although the decays are similar in shape, the exact form of the decays is different at each of the sample positions. Decays obtained at

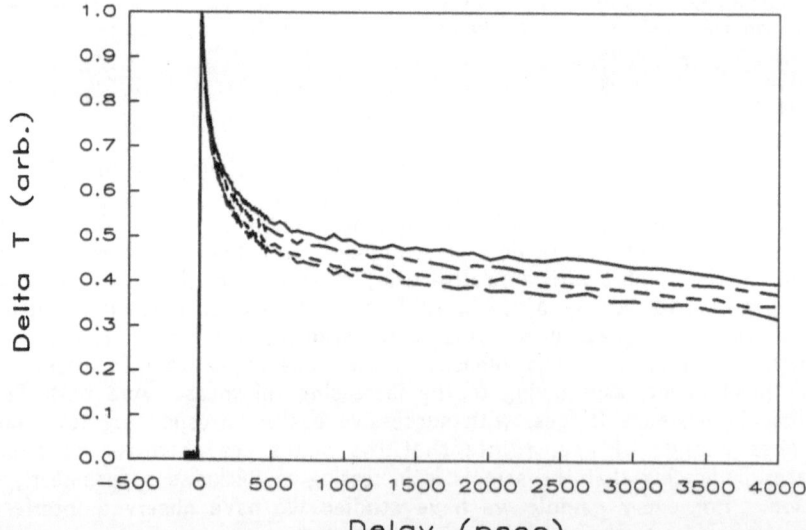

Figure 4. Decay of normalized ΔT at sample positions 265 (solid), 280 (- -), 287 (— —), and 300 (— - —). (See Fig. 1.) 2.11 eV pump; 2.02 eV probe; 2.4×10^{18} cm^{-3} carrier density.

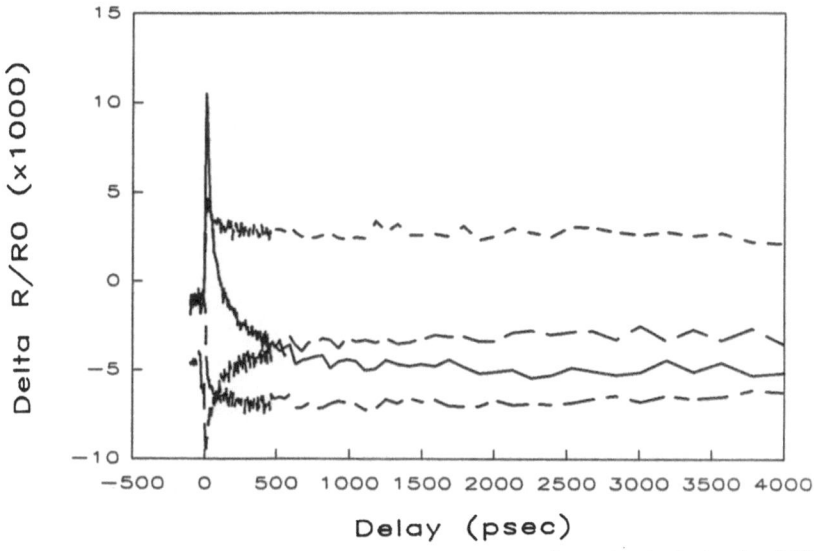

Figure 5. Decay of ΔR at sample positions 265 (solid), 280 (- -), 287 (— —), and 300 (— - —). (See Fig. 1.) 2.11 eV pump; 2.02 eV probe; 2.4×10^{18} cm^{-3} carrier density.

corresponding points along different interferogram cycles gave the same shapes. This suggests that the differences in the decays of ΔT were not due to sample inhomogenaities, but rather were related to an etalon effect.[30] In Figure 5 we compare the decays of ΔR at the same four sample positions. For comparison, in Figures 6 and 7 we show the decays of ΔT and ΔR, respectively, at different positions for a carrier density of 5.9×10^{17} cm^{-3}. We note that the same types of effects are observed, but at a lower signal level. At a given sample position the peak signal for the decay of ΔT is proportional to the initial carrier density (i.e. pump power).

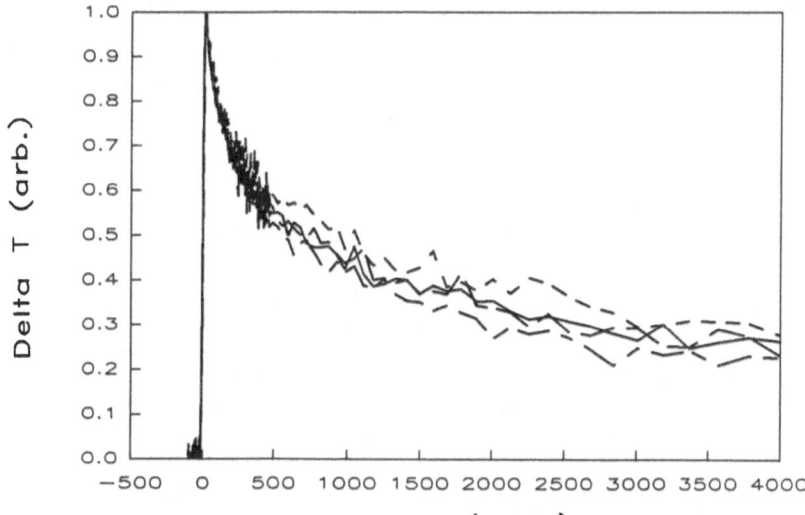

Figure 6. Decay of normalized ΔT at sample positions 240 (solid), 260 (- -), 270 (— —), and 280 (— - —). (See Fig. 1.) 2.11 eV pump; 2.02 eV probe; 5.9×10^{17} cm^{-3} carrier density.

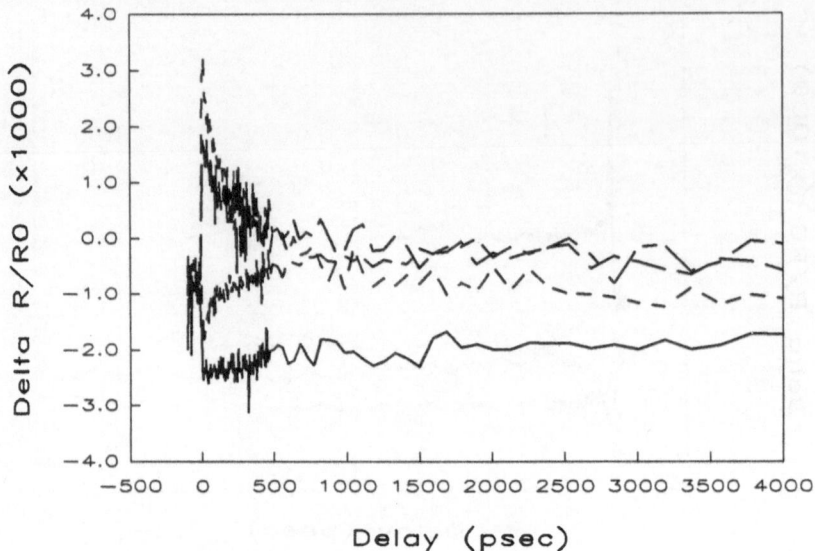

Figure 7.   Decay of ΔR at sample positions 240 (solid), 260 (- -), 270 (— —), and 280 (— - —). (See Fig. 1.) 2.11 eV pump; 2.02 eV probe; 5.9 x 10^{17} cm^{-3} carrier density.

For all the data obtained in this work we never observed a change in the decays with exposure time. At a given sample position and pump intensity the decay was the same regardless of the integrated light intensity. This is in contrast to results that are obtained with doped a-Si:H.[21,22]

Figure 8 shows the results of induced transmittance measurements as a function of sample position for various fixed delay times at a charge carrier density of 1.5 x 10^{18} cm^{-3}. The pump and probe wavelengths were 2.11 and 2.02 eV, respectively. Note that there are two effects present in the data: the decrease in peak *amplitude* with time, and the change in peak *position* with time. These effects are observed at all wavelengths and carrier densities that we have studied.[30]

DISCUSSION

The desired results of a picosecond PA experiment are to obtain the dynamics of an optical constant, preferably Δα; to model the response with respect to the mechanisms that contribute to charge carrier relaxation; and to correlate the data obtained for materials that have different electronic properties. In reviewing the data presented above the obvious concern is which of the PA decays, if any, represent the real physical processes that are ocurring in photoexcited a-Si:H. The results in Figure 8 give a pictorial explanation for the reason that we observe different decays at different positions in the interferogram. If the measurement begins at a position where ΔT is a maximum in the interference fringe the decay of ΔT will appear to be fast because the position of the peak shifts away from the original position to positions corresponding to a thicker sample. In a similar way, if the measurement begins at a position where ΔT is a minimum in the interference fringe the decay of ΔT will appear to be slow because the position of the peak shifts into the valley.

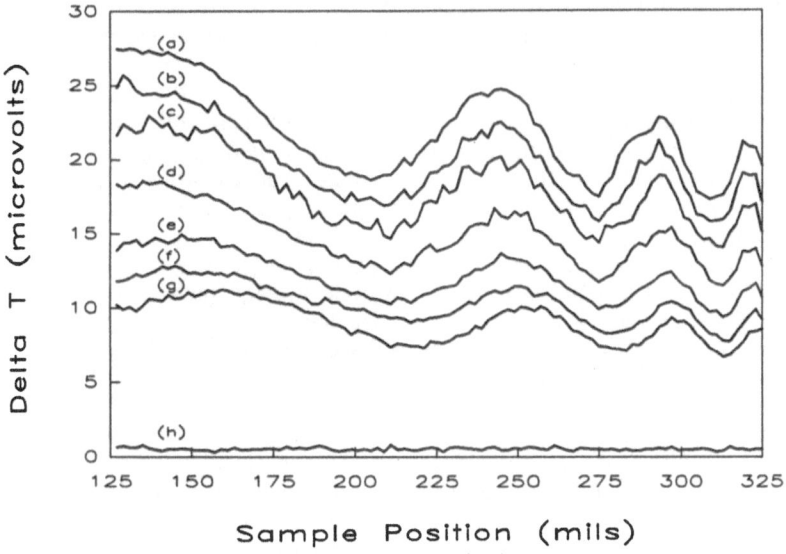

Figure 8. Intensity of ΔT as a function the sample position for delay times of (a) 20, (b) 40, (c) 83, (d) 164, (e) 351, (f) 1168, (g) 4000, and (h) -100 psec. 2.11 eV pump; 2.02 eV probe; 1.5×10^{18} cm^{-3} carrier density.

In order to understand the physical processes that contribute to the observed effects, we modeled ΔT using numerical differentiation of an equation for T_0 derived by Swanepoel.[33] The parameters in the equation are related to the measured changes in transmittance of light through the etalon formed by the sample, substrate system. We first modeled T_0 to obtain initial values for the absorption, α, index, n, and thickness, d. The thickness fit well to a parabolic function of the sample position only over a small region of the sample. We

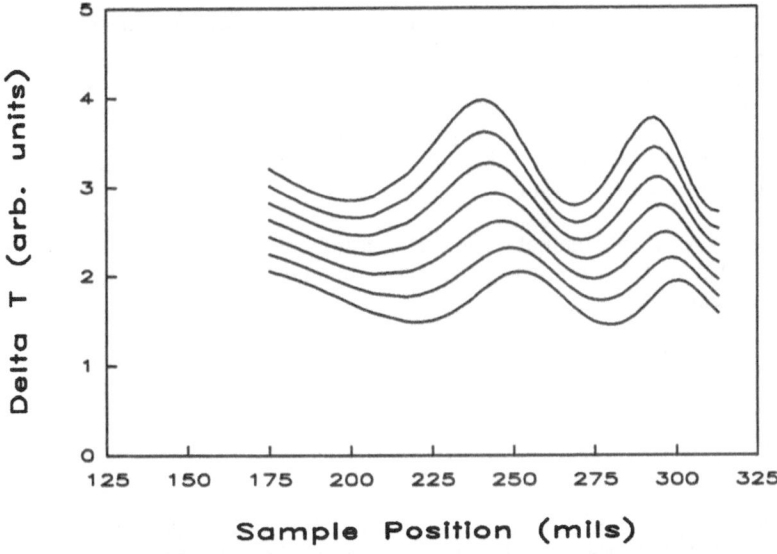

Figure 9. Model of the data presented in Fig. 8 obtained by numerical differentiation of an equation for the steady state transmittance. (See text for details.)

obtained values for α and n of 9500 cm^{-1} and 3.73, respectively. In Figure 9 we show the results of the numerical differentiation using incremental changes in α and n of 1.78 cm^{-1} and 4.0 x 10^{-5}, respectively. These parameters were chosen to provide a visual representation corresponding to the experimental data presented in Figure 8. The model accurately reproduces the important features of the data, including the decrease in the peak amplitude and the shift in the peak position. We found that the decay in the peak amplitude is related to the decay in induced absorption, $\Delta\alpha$; the change in peak position is related to an increase in the real part of the complex index of refraction, Δn. We note that without an incremental change in n the peak position does not change. The results of the modeling indicated that the cycle average (peak-to-peak integration) of ΔT is linear in $\Delta\alpha$. In a similar way, we have also modeled ΔR. Using the same parameters as we use to model ΔT we can reproduce the observed data for ΔR as a function of position.[34]

With this information we now can understand the reason that the decays are not the same at different positions. When the sample is photoexcited, both the real and imaginary parts of the complex index of refraction, **N**, change ($N = n + ik$; $k = \alpha\lambda/4\pi$); the change in the optical constants manifests itself in both ΔT and ΔR. Although the dominant contribution to ΔT is the change in the imaginary part of **N**, the real part of **N** distorts the ΔT decay (via the shift in the peak of the interferogram). The cycle average of ΔT as a function of position is linear in $\Delta\alpha$, which is the desired optical parameter.

The above discussion provides the basis for understanding the origin of the observed variations in the decay of ΔT and for providing an undistorted decay for comparison with other methods. However, the data acquisition for the cycle averaging is time consuming and the analysis is tedious. If we assume that n and k change only slightly, we obtain the following first order expressions for ΔT and ΔR:

$$\Delta T = (\partial T/\partial n)\Delta n + (\partial T/\partial k)\Delta k \qquad (1)$$

$$\Delta R = (\partial R/\partial n)\Delta n + (\partial R/\partial k)\Delta k \qquad (2)$$

The desired result is to obtain a decay that is only proportional to Δk (i.e. $\Delta\alpha$). Tauc and coworkers have recently shown that there are certain optical thicknesses in the sample where ΔT is directly proportional to Δk (i.e. At the maxima and minima of T_0 $(\partial T/\partial n) = 0$).[35] However, these positions must be determined very well and only slight changes in position will distort the decay curves. We have observed these positions experimentally, but practically they are difficult to locate *a priori*.

Another way to eliminate the contribution of Δn to the decay of ΔT is to combine ΔT and ΔR. Solving Eq. (1) for Δn and substituting the result in Eq. (2) we obtain

$$\Delta R - [(\partial R/\partial n)(\partial T/\partial n)^{-1}]\Delta T = [(\partial R/\partial k) - (\partial T/\partial k)(\partial R/\partial n)(\partial T/\partial n)^{-1}]\Delta k \qquad (3)$$

which is independent of Δn. Eq. (3) can be expressed as

$$\Delta R + f\Delta T = f'\Delta k \qquad (4)$$

where f and f' are functions of n, α, and d. We have evaluated numerically the various partial derivatives in the above equations in order to determine f and f'. As a function of optical thickness, f and f' oscillate with the same period but are 180° out of phase with T_0. This suggested the possibility that a linear

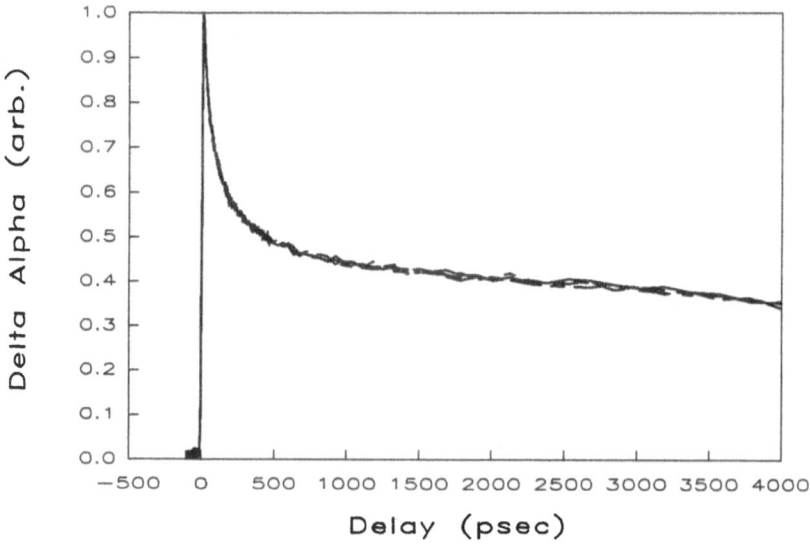

Figure 10.   Normalized sums of ΔT and ΔR for the decays shown in Figs. 4 and 5.   (See text for details.)

sum of ΔT and ΔR would give the same decay at all positions.   Indeed, we have found empirically that

$$\Delta R + c\Delta T \propto \Delta k \tag{5}$$

where $c = (R_{0,max} - R_{0,min})/(T_{0,max} - T_{0,min})$; the T_0 and R_0 maxima and minima are obtained from the curves in Figure 1.   (i.e. $R_0 + cT_0$ gives a curve free from interference fringes.)   Alternatively, c can be approximated to within ~10% by $c \sim 1/$(fraction of transmitted light).   This last approximation is valuable when the sample is flat and the optical constants are not known.   (We note that experimentally c must also be corrected for the relative sensitivity of the transmittance and reflectance detection systems.)

We used Eq. (5) to obtain position independent decays for the data presented previously.   The normalized sums of ΔT and ΔR for the curves from Figures 4 and 5 (carrier density of 2.4×10^{18} cm^{-3}) are shown in Figure 10. Within experimental error the four decays are superimposable.   We analyzed the data in Figures 6 and 7 (carrier density of 5.9×10^{17} cm^{-3}) in a similar way; the results are shown in Figure 11.   Again, within experimental error the four decays are superimposable.

In Figure 12 we plot on a log-log scale the results that we obtained from the cycle average of ΔT as a function of position and from the summation of the ΔT and ΔR decays for a carrier density of 5.9×10^{17} cm^{-3}.   We note that the agreement between the two sets of data is very good.   We have also verified that when we obtain a decay at a maxima or minima of T_0, where $(\partial T/\partial n) = 0$, the curve superimposes the curves in Figure 12.[35]   Based on our ability to reproduce the data using these different methods and to model the data, we are confident that the decay plotted in Figure 12 represents the undistorted decay of Δα.   For practical reasons it is easier either to locate accurately a maximum or minimum in T_0 and obtain the ΔT decay or to use the summation of the ΔT and ΔR decays in order to extract the parameters associated with the decay of Δα.   Destroying the etalon effect by roughening

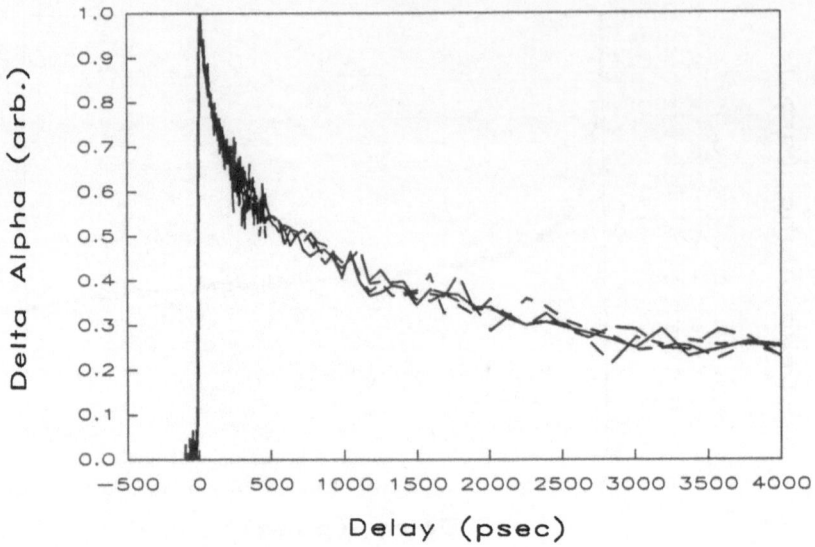

Figure 11.   Normalized sums of ΔT and ΔR for the decays shown in Figs. 6 and 7.   (See text for details.)

either the sample surface or the sample-substrate interface would also give undistorted decays.   This option is less desirable since it precludes the study of materials in most real devices and interfaces.

From our analysis of the decays obtained both by the cycle average of ΔT as a function of position and by the summation of the ΔT and ΔR decays, it is apparent that both the real and imaginary parts of **N** contribute to the decay of ΔT and the decay of ΔR when the sample, substrate system forms an etalon. This is to be expected since the Kramers-Kronig relation is valid for the photoexcited sample as well as the unexcited sample.   However, the manner in which the real and imaginary parts of the index of refraction contribute to the observed PA decays in a-Si:H has only recently been considered.[30,31,35]   Within our experimental error, Δα and Δn have the same dynamic behavior.   (Δn as a function of time can be determined from the shift in the peak position of the interferograms in Figure 8 or from the decay of ΔR at positions where $(\partial R/\partial k)$ = 0.[35])   Again, this is not unexpected since the changes in the electronic properties that give rise to Δα (i.e. photoexcited charge carriers) are the same changes that give rise to Δn.

With the confidence that our results represent the decay of Δα, we can consider the implications that our results have in terms of the mechanisms for carrier relaxation in a-Si:H.   Previous work by Tauc and coworkers[13,15] suggests that thermalization of the hot carriers occurs in ~1 ps.   Within the time resolution of our experiment we would probably not observe this effect.   Also, because the PA decay rates correlate with the density of uncompensated spins and because the observed temperature dependence of the decay contradicts the Onsager model, geminate recombination is considered to contribute negligibly to the PA decays.[15,19]   The dominate decay pathway at low carrier densities is considered to be monomolecular recombination through a trap or recombination center, specifically dangling bonds.[36]   At high carrier densities bimolecular recombination from tail states to tail states needs to be considered.   And for thin samples surface state recombination may be evident.[37]

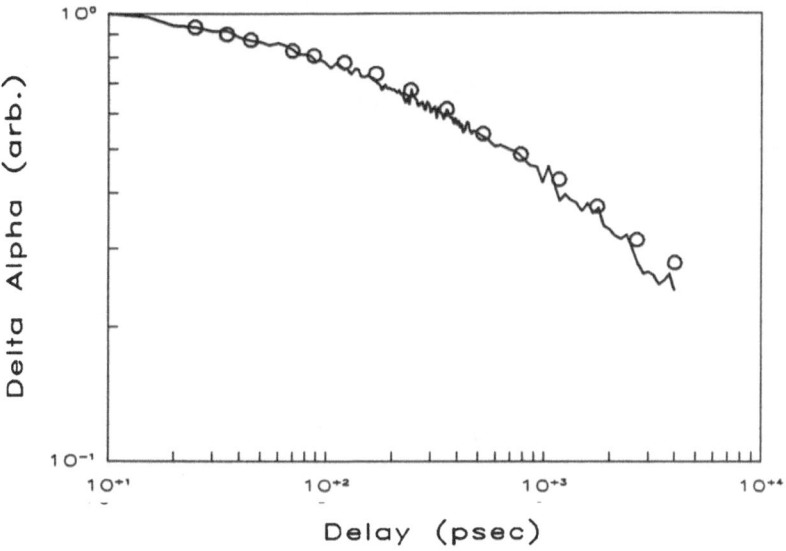

Figure 12. Comparison of the decays for $\Delta\alpha$ obtained from the cycle average of ΔT as a function of position (O) and from the summation of the ΔT and ΔR decays (solid) for a carrier density of 5.9 x 10^{17} cm^{-3}. (See text for details.)

A multiple trapping (MT) model[25-27] for nongeminate recombination of carriers has been used to explain previous PA decays in a-Si:H.[15,20] The model is expressed as follows:

$$N(t) = N(0)/[1 + (t/\tau)^\alpha] \qquad (6)$$

where τ is the mean trapping time constant and α is usually referred to as the dispersion parameter, first introduced by Scher and Montroll in their theory of dispersive transport.[23] (This use of α is not to be confused with the notation using α for denoting the absorption coefficient.) In the MT model α is temperature dependent. The long time limit of Eq. (6) is

$$N(t) = N'(0)t^{-\alpha} \qquad (7)$$

Eq. (6) is valid when the carrier transport is dominated by the interaction with shallow traps whose energy distribution is exponential.

In Figure 13 we compare on a log-log scale the sums of the decays that are shown in Figures 10 and 11. We also plot in Figure 13 the fit to Eq. (6) for the decay corresponding to a carrier density of 5.9 x 10^{17} cm^{-3}, and the fit to Eq. (7) for the decay corresponding to a carrier density of 2.4 x 10^{18} cm^{-3}. The fit for the lower carrier density gives τ = 440 ± 20 ps and α = 0.57 ± 0.01. These results compare with values of ~20 ps and ~0.7, respectively, obtained for phosphorous doped a-Si:H at room temperature.[15,20,21] These results indicate that the mean trapping time in intrinsic a-Si:H is longer than in the phosphorous doped material. This is consistent with the fact that the doped material has a greater defect density to which the carriers can relax.

The results that we obtain for the fit at the higher carrier density are more difficult to interpret. A fit to Eq. (7) gives α = ~0.20 for the decay. PL and PA studies[4,24] have shown that at carrier densities greater than 10^{18} cm^{-3}

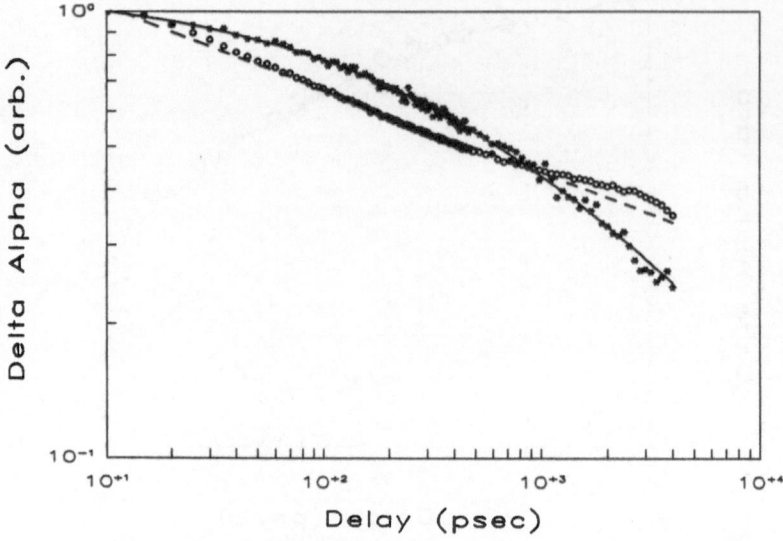

Figure 13. Comparison of the decays for $\Delta\alpha$ obtained from the summation of the ΔT and ΔR decays for a carrier density of 5.9×10^{17} cm^{-3} (*) and 2.4×10^{18} cm^{-3} (o). The solid and dashed lines are the fits to the data as explained in the text.

bimolecular recombination is an important loss mechanism for the carriers. However, contributions from bimolecular recombination would tend to increase α to values near 1.[15,24]

We need to consider if the changes in the decays with carrier density are associated with the peak carrier density, or if there is a steady state build-up of unrelaxed carriers owing to the high repetition rate of the laser source. In Table 1 we summarize the decay parameters obtained from a fit to Eq. (6) for decays obtained at two carrier densities for two repetition rates. If the decay depends only on the peak carrier density, the decay parameters should be independent of repetition rate. It is clear that both the peak carrier density and the steady state carrier density affect the observed PA decay. One possible explanation for these changes is that at higher light flux the primary relaxation pathways are saturated. These data suggest that models for carrier relaxation based on PA decays must consider carefully the incident pump intensity and the repetition rate of the excitation source.

SUMMARY

The most significant observation from our results is the fact that PA decays for most studies of a-Si:H have the potential for being distorted owing to the way that the real and imaginary parts of the complex index of refraction contribute to the induced transmittance and the induced reflectance. We have presented several methods for alleviating the problem and obtaining decays that represent the change in the induced absorption. We have shown that the decay of the induced absorption depends strongly on the peak carrier density and the repetition rate of the excitation source. The PA data can be fit to the model for multiple trapping. In future work we plan to obtain the temperature dependence of the decay parameters and study the wavelength dependence of the induced absorption decay.

Table 1. Comparison of the decay parameters obtained at different carrier densities and source repetition rates.

Peak Carrier Density (cm^{-3})	Repetition Rate	
	0.5 MHz	1.0 MHz
5.7×10^{17}	$\tau = 400 \pm 10$ ps $a = 0.58 \pm 0.01$	$\tau = 170 \pm 10$ ps $a = 0.45 \pm 0.01$
2.1×10^{18}	$\tau = 11 \pm 1$ ps $a = 0.35 \pm 0.01$	$\tau = 1.5 \pm 0.2$ ps $a = 0.29 \pm 0.01$

ACKNOWLEDGEMENTS

We want to thank R. W. Collins and H. Scher for their many thoughtful contributions to this work. We appreciate the help of D. A. Chernoff, H. L. Fang, and R. L. Swofford. We also acknowledge helpful conversations with T. E. Orlowski and Z. Vardeny. We are particularly grateful to J. Tauc for providing us with a preprint of his work. (Reference 35.)

REFERENCES

1. A. M. Johnson, D. H. Auston, P. R. Smith, J. C. Bean, J. P. Harbison, and A. C. Adams, Phys. Rev. B 23, 6816(1981).
2. T. E. Orlowski, Solid State Commun. 56, 265(1985).
3. T. E. Orlowski, Bull. Am. Phys. Soc. 28, 239(1983).
4. C. Tsang and R. A. Street, Phys. Rev. B 19, 3027(1979).
5. R. W. Collins, P. Viktorovitch, R. L. Weisfield, and W. Paul, Phys. Rev. B 26, 6643(1982).
6. B. Rauscher and B. Bullemer, Phys. Stat. Sol. (a) 79, 161(1983).
7. B. A. Wilson, P. Hu, T. M. Jedju, and J. P. Harbison, Phys. Rev. B 28, 5901(1983).
8. T. E. Orlowski and B. A. Weinstein, Phil. Mag. B 52, 1(1985).
9. B. A. Wilson, T. P. Kerwin, and J. P. Harbison, Phys. Rev. B 31, 7953(1985).
10. D. G. Stearns, Phys. Rev. B 30, 6000(1984).
11. D. J. Dunstan and F. Boulitrop, Phys. Rev. B 30, 5945(1984).
12. R. A. Street, D. K. Biegelsen, and R. L. Weisfield, Phys. Rev. B 30, 5861(1984).
13. Z. Vardeny and J. Tauc, Phys. Rev. Lett. 46, 1223(1981).
14. Z. Vardeny and J. Tauc, Opt. Commun. 39, 396(1981).
15. J. Tauc, Hydrogented Amorphous Silicon, Part B, J. Pankove, Ed., (Academic Press, New York, 1984), pp. 299.
16. Z. Vardeny, J. Non-Cryst. Solids, 59 & 60, 317(1983).

17. D. E. Ackley, J. Tauc, and W. Paul, Phys. Rev. Lett. 43, 715(1979).
18. J. Kuhl, E. O. Gobel, Th. Pfeiffer, and A. Jonietz, Appl. Phys. A 34, 105(1984).
19. Z. Vardeny, J. Strait, and J. Tauc, Appl. Phys. Lett. 42, 580(1983).
20. Z. Vardeny, J. Strait, D. Pfost, J. Tauc, and B. Abeles, Phys. Rev. Lett. 48, 1132(1982).
21. J. Strait, Ph.D. Thesis, (Brown University, May, 1985), pp. 75.
22. J. Strait and J. Tauc, Appl. Phys. Lett. 47, 589(1985).
23. H. Scher and E. W. Montroll, Phys. Rev. B 12, 2455(1975).
24. Z. Vardeny, P. O'Connor, S. Ray, and J. Tauc, Phys. Rev. Lett. 44, 1267(1980).
25. H. Scher, J. Phys., Paris 42, 547(1981).
26. T. Tiedje and A. Rose, Sol. State Commun. 37, 49(1980).
27. J. Orenstein and M. Kastner, Phys. Rev. Lett. 46, 1421(1981); J. Orenstein and M. Kastner, Sol. State Commun. 40, 85(1981).
28. D. M. Roberts, J. F. Palmer, and T. L. Gustafson, *Spectroscopic Characterization Techniques for Semiconductor Technology II*, F. H. Pollak, Ed., Proc. SPIE 524, 106(1985).
29. D. M. Roberts and T. L. Gustafson, Opt. Commun. 56, 334(1986).
30. D. M. Roberts and T. L. Gustafson, J. Non-Cryst. Solids 77 & 78, 551(1985).
31. D. M. Roberts and T. L. Gustafson, *Amorphous Semiconductors for Microelectronics*, D. Adler, Ed., Proc. SPIE 617, xxx(1986).
32. R. W. Collins and W. J. Biter, Optical Effects in Amorphous Semiconductors, Snowbird, Utah, 1984, P. C. Taylor and S. G. Bishop, Ed., (AIP Conference Proceedings #120, AIP, New York, 1984), pp. 170.
33. R. Swanepoel, J. Phys. E: Sci. Instrum. 16, 1214(1983).
34. D. M. Roberts and T. L. Gustafson, unpublished results.
35. H. T. Grahn, C. Thomsen, and J. Tauc, Opt. Commun., in press.
36. R. S. Crandall, Hydrogented Amorphous Silicon, Part B, J. Pankove, Ed., (Academic Press, New York, 1984), pp. 245.
37. V. J. Newell, T. S. Rose, and M. D. Fayer, Phys. Rev. B 32, 8035(1985).

PICOSECOND TRANSIENT GRATING EXPERIMENTS ON HYDROGENATED AMORPHOUS SILICON:

A MODEL FOR SURFACE QUENCHING

Vincent J. Newell and M. D. Fayer

Department of Chemistry
Stanford University
Stanford, California 94305

I. INTRODUCTION

In this article we present a room temperature experimental study of
the transport and surface quenching of optically generated charge
carriers in thin-film hydrogenated amorphous silicon (a-Si:H). At low
temperature (4.2 K), in high quality samples, the principal mechanism for
carrier quenching is radiative recombination.[1] Nonradiative processes
such as Auger recombination and tunneling to defects play only a minor
role.[1] However, it has been suggested[2] that surface effects play an
important role at higher temperatures (77 K). Rehm et al.[2a] examined the
effects of sample thickness and wavelength (optical penetration depth) on
the decay time of the luminescence. They found dramatic effects when the
sample thickness or the optical penetration depth fell below 0.3 μm.

In the experiments presented below, we use a picosecond transient
grating technique[3] to investigate carrier transport in 0.33 μm and 0.16
μm samples of a-Si:H. The application of the picosecond transient
grating technique is based on earlier investigations of excitation[4] and
exciton[5] transport in molecular crystals. The transient grating
experiment works in the following manner (see fig. 1a). A picosecond
time scale pulse of light is split in two. The paths of the resulting
pulses are arranged to have a known angle between them and to intersect
simultaneously in the sample. Interference between the two coherently
related pulses creates an optical fringe pattern in the sample such that
the intensity of light varies sinusoidally in the beam overlap region.
The interference fringe spacing is determined by the angle between the
beams and by the wavelength of the light.

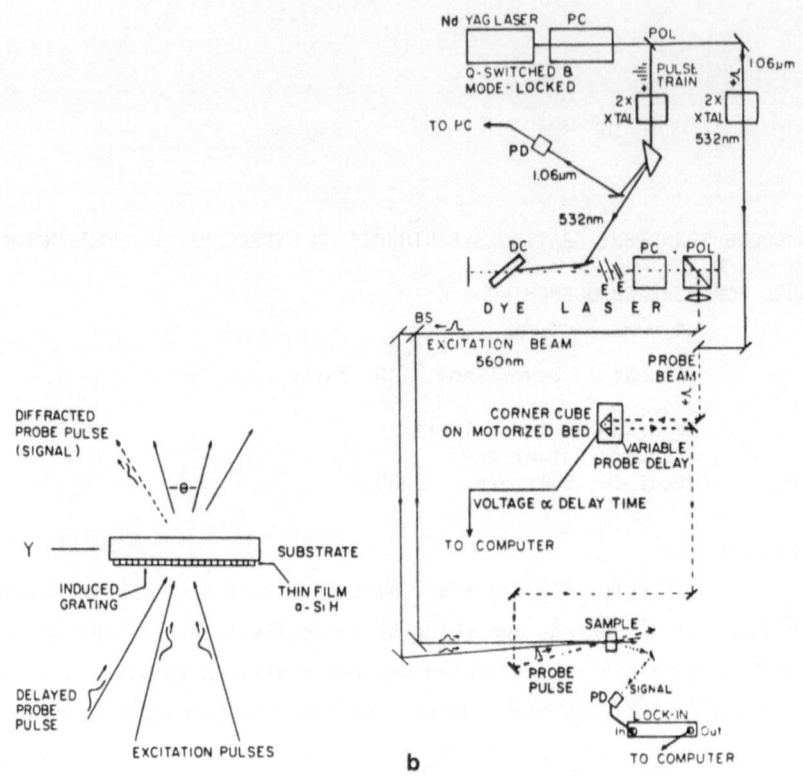

Fig. 1. (a) Transient grating pulse seqkuence (see text for description).

(b) The transient grating experimental set-up. DC = dye cell, POL = polarizer, PC = Pockels cell, BS = beamsplitter, E = etalon, PD = photodiode (see text for description).

The wavelength of the excitation pulses is tuned above the bandgap. Electron-hole pairs are excited and relax to band edge localized states. The relaxation occurs on a time scale fast compared to the time required for transport over macroscopic distances.[6] The localized carriers will have the same spatial distribution as the sinusoidal optical interference pattern, i.e., there will be a continuous oscillatory variation in the concentration of electron-hole pairs. After a suitable time delay, a probe pulse (which may differ in wavelength from the exciting pulses) is directed into the sample along a third path. The probe pulse will experience an inhomogeneous optical medium resulting from alternating regions of high and low concentrations of electron-hole pairs. The separated charges associated with an electron-hole pair perturbs the electron distribution in the local environment, and therefore changes the index of refraction. The result is that the spatial variation in the number density of electron-hole pairs produces a spatially periodic variation in the index of refraction. Thus, the probe pulse encounters a diffraction grating (phase grating) which causes it to diffract into one or more orders (see fig. 1a). Phase gratings have been previously observed in experiments on crystalline silicon.[7] The diffracted pulse leaves the sample along a unique direction.

In the experiments described below, the grating wave vector is in the plane of the a-Si:H sample. For small fringe spacing ($\sim$ 1 μm) carrier transport dominates the grating decay. Migration will move carriers from areas of high concentration (grating peaks) to areas of low concentration (grating nulls). Thus, the motion will fill in the grating nulls and deplete the peaks. Destruction of the grating pattern by spatial redistribution of the carriers over macroscopic distances will lead to a decrease in the intensity of the diffracted probe pulse as the probe delay time is increased. For a small fringe spacing, time dependence of the grating signal is directly determined by the rate of transport. As the fringe spacing is increased, the decay of the signal arising from spatial transport is slowed because it takes longer for carriers to move from grating peaks to grating nulls.

For sufficiently large fringe spacing, the decay of the grating due to in-plane transport will become negligible compared to other decay pathways. The large fringe spacing time dependent signal is independent of fringe spacing; therefore, the dynamics of other processes can be investigated. In experiments involving electronic excited state gratings, the large fringe spacing decay is due to the exponential decay of the excited states,[3-5] and the transient grating signal decay is a single exponential. In the experiments described below, the large fringe spacing decay is highly nonexponential and depends on the thickness of the thin-film. A model is presented which relates the large fringe spacing decay to carrier diffusion and quenching at the thin-film surfaces. The model, which assumes the surface recombination rate to be infinite, provides a good description of the nonexponential time dependence and thickness dependence of the signal in terms of a diffusion constant $D_\perp$ for transport perpendicular to the plane of the thin-film. $D_\perp$ is distinct from the diffusion constant $D_\parallel$ measured in the thin-film plane at small fringe spacing. We find $D_\parallel = 1.0 \times 10^{-2}$ cm^2/sec and $D_\perp = 0.4 \times 10^{-2}$ cm$_2$/sec. The results demonstrate that surface quenching is the dominant decay path for photo-generated carriers at room temperature and that the surface recombination occurs on the first encounter with the surface region on a time scale that is fast compared to the rate of diffusion to the surface.

II. Mathematical Formulation

First consider a system in which the photogenerated carriers undergo diffusive transport but do not decay either by surface quenching or other recombination processes. The optical interference pattern produces electron-hole pairs in the medium such that the initial spatial distribution along the grating axis (y direction) is given by[3-5]

$$N(y,0) = 1/2(1 + \cos\Delta y) \tag{1}$$

where

$$\Delta = 2\pi/d \tag{2}$$

and d, the grating fringe spacing, is given by

$$d = \frac{\lambda}{2 \sin(\theta/2)} \tag{3}$$

In eq. (3) λ is the wavelength of the excitation pulses (in air) and θ is the angle between them (in air) (see fig. 1a). A variable time delay probe pulse is diffracted off the induced grating. The diffracted probe is monitored as a function of time delay between the probe and the excitation pulses.

The change in signal, as a function of time (i.e., the decay of the grating) results from motion of the carriers along the grating axis. By solving the appropriate diffusion equation for the initial grating condition established at $t = 0$ the time-dependent signal has been shown to be[8,9]

$$S(t,\theta) = Ae^{-Kt} \tag{4}$$

where A, a time independent constant, describes the strength of the signal and depends on beam geometries, laser intensities, and other experimental factors. For small θ, the decay constant, K, is given by

$$K = 2\Delta^2 D_\parallel = (8\pi^2/\lambda^2)\theta^2 D_\parallel. \tag{5}$$

It is seen that for diffusive transport along the grating axis, the signal decays exponentially at a rate which depends directly on the diffusion constant in the grating wave vector direction. In the absence of carrier decay processes the diffusion constant can be obtained from one grating decay measurement at a particular value of θ. In practice, however, it is generally better to evaluate the decay as a function θ. If transport is responsible for the grating decay, then a plot of K vs. θ^2 should yield a straight line. The diffusion constant is obtained from the slope of the line. This is the procedure employed below.

In real systems, the grating decays due to processes other than diffusion along the grating wave vector direction. The time dependence of the grating arising from these other processes is independent of θ. The total signal can be written as

$$S(t,\theta) = Ae^{-Kt}f(t). \tag{6}$$

In the limit that θ is very small (a large fringe spacing), the exponential term is constant and the signal becomes angle independent

$$S(t,\theta \approx 0) = S_0(t) = Af(t). \tag{7}$$

For excited state gratings previously reported, the small angle decay is dominated by the single exponential decay of the excited states[3,9a]; $f(t) = \exp(-2t/\tau)$ where τ is the excited state lifetime. Here we will present a model in which $f(t)$ is not exponential and is determined by surface quenching of the electron-hole pairs. Regardless of the form of $f(t)$, the angle dependent part of the signal, $S_\theta(t,\theta)$, which yields the in-plane diffusion constant, $D_\parallel$, can be obtained by dividing eq. (6) by eq. (7), i.e.,

$$S_\theta(t,\theta) = \frac{S(t,\theta)}{S_0(t)} = e^{-Kt}. \tag{8}$$

At large fringe spacing, the signal is fringe spacing independent. However, the signal is nonexponential and is dependent on the sample thickness. To account for this we present a model which allows us to calculate the time dependent probability that carriers initially located at various distances from the samples surface or interface, have interacted with either of the sample boundaries. The sample boundaries are located at $x = 0$ and $x = s$. The initial distribution of electron-hole pairs is given by Beer's law.

The probability that carriers originating at x_0 have quenched by diffusing to one of the boundaries at $x = 0$ and $x = s$ is[8]

$$P_Q(x_0,t) = \frac{1}{2}\left[\text{erfc}\left(-\frac{x_0}{2\sqrt{D_\perp t}}\right) + \text{erfc}\left(\frac{s-x_0}{2\sqrt{D_\perp t}}\right)\right] \tag{9}$$

where erfc is the error functions complement. The probability that carriers with an initial distribution throughout the depth of the sample, $f(x_0)$, have reached one of the boundaries is

$$P_Q(t) = \frac{1}{2}\int_0^s f(x_0)\left[\text{erfc}\left\{-\frac{x_0}{2\sqrt{D_\perp t}}\right\} + \text{erfc}\left\{\frac{s-x_0}{2\sqrt{D_\perp t}}\right\}\right]dx_0 \tag{10}$$

The distribution function is normalized between the boundaries 0 and s. For a Beer's law initial distribution assumed here,

$$f(x_0) = \frac{\alpha e^{-\alpha x_0}}{(1 - e^{-\alpha s})} \qquad (11)$$

where α is the absorption coefficient.

The time development of the transient grating signal, $S_0(t)$, is proportional to the square of the peak-null difference in the density of carriers.[9] In the large fringe spacing case, as described by this model, the number of carriers in the nulls remain constant in time, and for equal amplitude excitation pulses the constant is equal to zero. Therefore, the signal is proportional to the square of the carriers remaining in the sample. The surface recombination rate is considered to be infinitely fast on the time scale of the experiment and the signal is given by[8]

$$S_0(t) = A[1 - P_Q(t)]^2. \qquad (12)$$

The model displays a nonexponential time decay which is dependent on the sample thickness provided s is on the order of the Beer's length, $1/\alpha$. For $s \gg 1/\alpha$, the carriers are generated so far from the back surface of the sample that the signal decay at relatively short times is completely dominated by quenching at the front surface. However, when s $\approx 1/\alpha$, the time dependence is a sensitive function of s. This behavior is displayed in the data reported below.

The model described above calculates the probability of interacting with the surface as the fraction of spreading Gaussians which is past the interfaces. Actually, this is a first passage problem, i.e., the time dependent probability that carriers encounter the surface for the first time should be calculated. However, the difference between the procedure used here and a first passage calculation will be small.

III. Experimental Procedures

The experimental system is illustrated schematically in fig. 1b. The laser is a cw pumped, acousto-optically Q-switched and mode-locked Nd:YAG system which produces 1.06 μm pulse trains at 400 Hz.[10] A single pulse of 80 ps duration and ~ 40 μJ in energy is selected by a Pockels cell. The selected IR pulse is then frequency doubled by a CD*A crystal to produce a 532 nm, 60 ps, 15 μJ Tem$_{oo}$ pulse. The remaining 1.06 μm pulse train is also frequency doubled and used to synchronously pump a dye laser which produces a 560 nm, 30 ps, 10 μJ dye laser pulse.[9a]

The 560 nm single pulse is passed through a 50% beamsplitter and the resulting two pulses are recombined to form the interference pattern at the sample. The 532 nm single pulse is used as the probe. A retroreflector is drawn along a precision optical rail by a motor which provides continuous scanning of the probe delay. A ten-turn potentiometer, also driven by the motor, provides a voltage proportional to the probe delay. The diffracted signal is spatially and spectrally filtered and is detected by a large area photodiode and lock-in amplifier. The output of the lock-in amplifier and the voltage proportional to the probe time delay are digitalized and stored by a computer for subsequent analysis.

The a-Si:H samples were provided by Dr. R. D. Wieting of Arco Solar Corporation. The samples were prepared by standard glow discharge procedures on suprasil substrates. The substrate temperature was held at 200 °C. A silane pressure of 1 Torr was used, and the growth rate was a few Å/sec. Several substrates were coated in each run. The results presented below did not vary for different samples from the same coating run, nor did they vary for different spots on a given sample. Samples with different thicknesses show significant differences, as discussed in the next section. The samples were mounted in air. Spot sizes were 300 μm for the excitation beams and 230 μm for the probe beam. The beams were attenuated to approximately 2 μJ per beam. For the power levels employed to obtain the results presented below, no damage to the samples occurred during the tens of minutes required to conduct a single experiment.

IV. Results and Discussion

Figure 2 shows the time dependent decay of the transient grating signal for two different fringe spacings. The sample is 0.33 μm thick. The top trace is for a fringe spacing of 6.3 μm. For fringe spacings larger than 5 μm the grating decay is fringe spacing independent. This decay defines $S_0(t)$ (eq. (7)) for this sample. The lower trace is for a fringe spacing of 1.0 μm. This decay is substantially faster than the upper trace due to carrier transport from grating peaks to grating nulls. The function, $S_\theta(t,\theta)$, which characterizes the in-plane transport ($D_\parallel$) is obtained from the data using eq. (8). Figure 3a shows a plot of K vs. θ^2. The points fall on a straight line and from eq. (5), yield a diffusion constant, $D_\parallel = 1.0 \times 10^{-2}$ cm^2/sec.

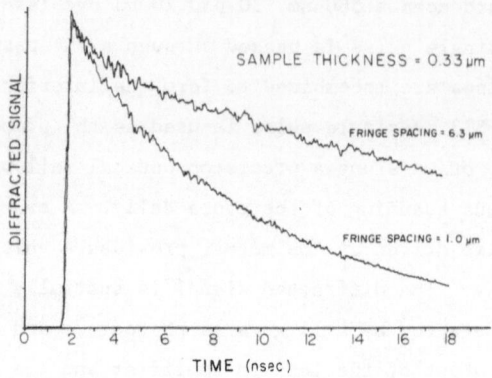

Fig. 2. Transient grating signal decay for the 0.33 μm thick
sample at two different fringe spacings. The upper
decay curve at 6.3 μm fringe spacing is highly nonexponential and is dominated by surface quenching of
the photo-excited carriers. The decay for 6.3 μm fringe
spacing is in the large fringe spacing limit where the
shape of the decay is determined by sample thickness
and is insensitive to the fringe spacing. The lower
decay curve at 1.0 μm fringe spacing is faster due to
diffusion of carriers in the plane of the sample from
the grating peaks to the nulls which destroys the
grating pattern.

This data was reproducible for various spots on the same sample and
various samples made in the same deposition run, provided the excitation
and probe pulses were sufficiently attenuated. With the full laser power
and ~ 300 μm spot sizes, permanent gratings were burned into the sample.
After excitation with full power, extremely intense, time independent,
diffraction occurred even when the excitation beams were blocked.
Examination of the sample showed that a striped pattern had been
produced. It appeared as if the a-Si:H had been removed from the sample.
In the data presented, the beams were attenuated until no damage of the
samples occurred. The decay curves were also checked to see if the time
dependence was affected by power. For the energies used in the
experiments, the data was power independent. The effect of repetition
rate was also tested, and the decays were found to be independent of
repetition rate. This indicates the absence of long lived species (> 1
msec) which are important at very low temperatures.[1]

Figure 3b shows a plot of K vs. θ^2 for a 0.16 μm sample. Again the
data falls on a straight line and yields a value of $D_\parallel$ ~ 0.8×10^{-2}
cm^2/sec. While this is very close to the $D_\parallel$ obtained for the thicker

232

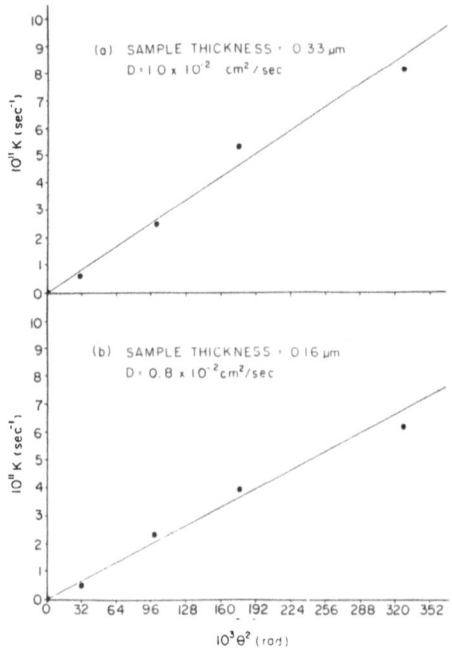

Fig. 3. (a) K vs. θ^2 for a 0.33 μm thick a-Si:H film using eq. (11). For diffusive transport, the points should fall on a straight line. The slope equals $8\pi^2 D_{\parallel}/\lambda^2$ and therefore yields the value of the diffusion constant, $D_{\parallel} \sim 1.0 \times 10^{-2}$ cm^2/sec; (b) K vs. θ^2 for the 0.16 μm thick film. For this sample $D_{\parallel} \sim 0.8 \times 10^{-2}$ cm^2/sec.

sample, it falls somewhat outside of the range of experimental uncertainty. Whether this difference is due to the different sample thicknesses or arises because the samples were made in different deposition runs is presently unclear.

Figures 4a and 4b show decay curves taken on the 0.33 μm and 0.16 μm samples respectively with a 6.3 μm fringe spacing. At this fringe spacing, the decays are fringe spacing independent. The decays in the two samples are clearly different. These decays are nonexponential. The solid lines through the data were calculated using eq. (12). The data in fig. 4a was fit with one adjustable parameter, $D_{\perp}$. The best value for $D_{\perp}$ = 0.4 × 10^{-2} cm^2/sec. The solid curve through the data in fig. 4b was calculated without adjustable parameters using the $D_{\perp}$ value obtained from the fit in fig. 4a.

The model explains both the nonexponential shapes of the curves and the change in shape with sample thickness for the large fringe spacing decay. Electron-hole pairs migrate to the vicinity of the surface where

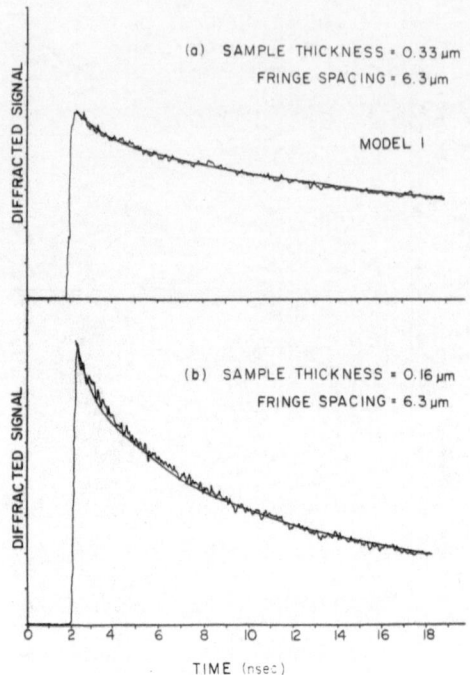

Fig. 4. The data in 4(a) and 4(b) are for the same fringe spacing
(6.3 μm) but different sample thickness, 0.33 and 0.16 μm,
respectively. The decay in the thicker sample is highly
nonexponential and substantially slower than the decay in
the thinner sample. The solid curve in (a) was fitted to
data using eq. (12) and one adjustable parameter, $D_\perp$.
A value of $D_\perp = 0.4 \times 10^{-2}$ cm^2/sec gave the best fit.
The solid curve in (b) was calculated without adjustable
parameters using the $D_\perp$ value obtained from the data in
(a). This demonstrates that the model is in good agreement
with the data.

they undergo a radiationless recombination process. The nonexponential
form of the decays is due to the distribution of transit times from the
bulk of the sample to the surfaces. The thicker sample has a broader
distribution of transit times than the thinner sample, resulting in a
differently shaped decay curve. The quantitative agreement between the
data and the calculated curves is strong support for the basic idea that
surface quenching is the dominant carrier recombination process in thin-
film a-Si:H at room temperature and that the surface recombination rate
is effectively infinite on the time scale of the experiment. The fact
that $D_\perp$ is different from $D_\parallel$ could be due to sample inhomogeneities in
the film growth direction ($\perp$ direction).

V. Conclusion

Using a picosecond transient grating technique we have obtained information on the dynamics of photogenerated carriers in thin-film a-Si:H on a nanosecond time scale. The diffusion constant, $D_\parallel$, for transport in the film plane was obtained from the fringe spacing dependence of the signal decays. The nonexponential shape of the decays and their dependence on sample thickness was explained by a model based on surface quenching. The diffusion constant, $D_\perp$, for transport perpendicular to the film plane was obtained from the data fits. In the model photoexcited carriers migrate to a surface state and recombine at rates that are infinitely fast on the time scale of the experiment. Good data fits were obtained with a value of $D_\perp = 0.4 \times 10^{-2}$ cm^2/sec. Recent calculations which assume an isotropic diffusion constant ($D_\perp = D_\parallel$) but finite unimolecular and bimolecular recombination rates have been performed. Preliminary indications are that these other models cannot fit the data. These results will be presented in a future publication.

The macroscopic surface quenching model in eq. (12) works because it embodies the essential features of the physical situation. There is a distribution of starting locations in the bulk of the material. This results in a wide distribution of transit times to the surface. The transit time to the surface is the rate limiting step while the surface recombination rate is effectively infinite.

Acknowledgment

We would like to thank professor Nathan Lewis for many valuable dis-cussions pertaining to surface quenching in these experiments. This work was supported by the National Science Foundation, Division of Materials Research (DMR84-16343) and by Department of Energy, Office of Basic Energy Sciences (DE-FG03-84ER13251).

References

1. R. A. Street, Adv. Phys. Adv. Phys. 30(5), 593 (1981);
    R. A. Street in: Semiconductors and semimetals, Vol. 21, Part B,
    ed. J. I. Pankove (Academic Press Inc., 1984) p. 197.
2. a) W. Rehm, R. Fischer, and J. Beichler, Appl. Phys. Lett. 37(5),
       445 (1980);
    b) R. A. Street and D. K. Biegelsen in: Topics in Applied Physics,
       Vol. 56, The Physics of Hydrogenated Amorphous Silicon II,
       ed. J. D. Joannopoulos and G. Lucovsky (Springer-Verlag, 1984)
       p. 195; R. A. Street and J. C. Knights, Phil. Mag. B. 43(6),
       1091 (1981).

3. H. J. Eichler, Optica Acta **246**, 631 (1977).

4. J. R. Salcedo, A. E. Siegman, D. D. Dlott, and M. D. Fayer, Phys. Ref. Letters **41**, 131 (1978).

5. T. S. Rose, R. Righini, and M. D. Fayer, Chem. Phys. Letters **106(1,2)**, 13 (1984).

6. E. O. Gobel and W. Grandszus, Phys. Rev. Lett., **48**, 1277 (1982); Z. Vardeny, J. Strait, and J. Tauc, in: Picosecond Phenomena III, Eds. K. B. Eisenthal, R. M. Hochstrasser, W. Kaiser, and A. Laubereau (Springer-Verlag, Berlin, 1982) p. 372.

7. J. P. Woerdman, Philip. Res. Repts. Suppl. No. 7 (1971).

8. V. J. Newell, T. S. Rose, and M. D. Fayer, Phys. Rev. B **32(12)**, 8035 (1985).

9a. K. A. Nelson, R. Casalegno, R. J. D. Miller, and M. D. Fayer, J. Chem. Phys. **77**, 1144 (1982).

9b. R. F. Loring and S. Mukamel, J. Chem. Phys. **83**, 1 (1985).

10. D. J. Kuizenga, D. W. Phillion, T. Lund, and A. E. Siegman, Opt. Commun. **9**, 221 (1973).

LANGMUIR-BLODGETT ELECTRONIC DEVICES

Scott E. Rickert, Jerome B. Lando, and Clifford D. Fung

Department of Macromolecular Science
Case Western Reserve University
Cleveland, Ohio 44106

ABSTRACT

The importance of control over the micro-engineering of Langmuir films
is amply documented for a variety of amphiphilic materials. Such parameters
as temperature, pressure, substrate quality, amphiphile chemistry, and
others must be considered when high-quality films are desired. A comparison
is made between high and low-quality films, and a novel MLSFET device is
described, which was prepared with a high-quality film. They are the initial products of a joint Polymer Science and Electrical Engineering laboratory established in late 1983, the Polymer Microdevice Laboratory (PML).
These devices exhibit characteristics well within the useful range of normal semiconductors, including temperature stability. Several of these devices will be discussed, along with the peculiar problems associated with
the fabrication and testing of such devices.

Many devices were prepared on a special 2 inch silicon wafer, upon
which was impregnated the 400 integrated devices, consisting of Field-Effect
Transistors (FET's) of varying gate geometries, interdigitated electrodes
for resistance and conductance measurements, and capacitance measurements.
The wafer is unique, in that all furnace steps were implemented prior to
the application of the multilayer to the silicon surface. The multilayer,
while stable up to 125°C, could not stand the very high temperatures of
the oxidation furnace. However, it could withstand the aluminum metalliza-
tion and plasma etching steps required, with no appreciable degradation.

INTRODUCTION

Although many researchers have investigated the potential applications
of Langmuir films (i.e. monolayers and multilayers), including those applications relating to electron devices, few have focused on those problems
which must be solved before such devices can become practical. The PML
is dedicated to solving the processing, and related quality-control problems
associated with Langmuir film fabrication.

These problems are more related to engineering than to science, but a
multi-disciplinary approach involving engineers and scientists is necessary
in order to solve many of them. Due to the nature of the problems, it
is necessary to focus on several simple amphiphilic systems, in order to
reduce the organic synthesis of the work.

PARAMETER	CONDITION
Subphase	'aged' water or low concentration of $CdCl_2$
Solution Concentration	0.5 – 1 mg/ml
Spreading Area	0.6 square nanometer/molecule
Dwell Period	10 minute
Compression Rate	2 cm/min
Dipping Speed	1 – 3 mm/min

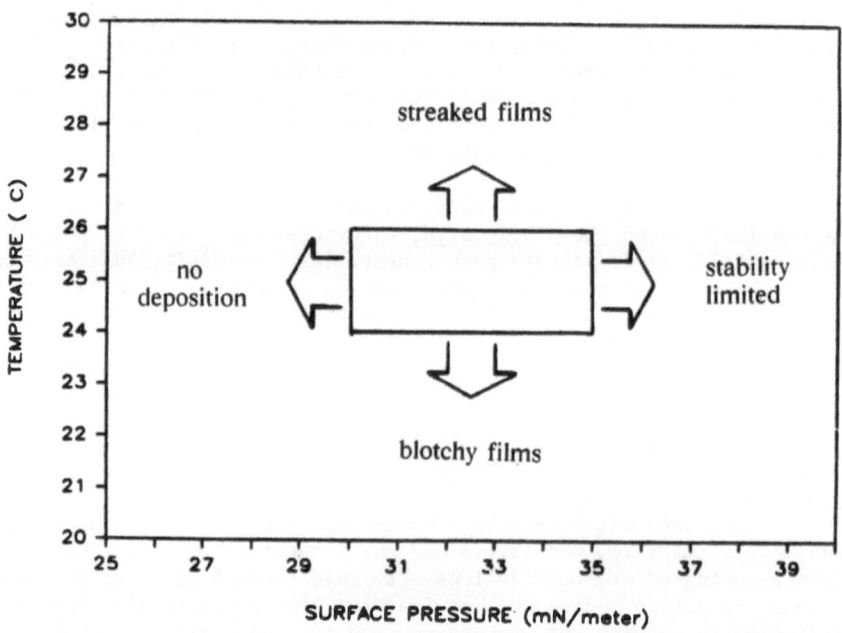

Figure 1. A typical Langmuir-Blodgett "processing window."

EXPERIMENTAL

A variety of amphiphiles, including stearic acid, brassidic acid, elaidic acid, and 16,8-diacetylene acid (168DA), were purified in a class 100 ultra-clean facility, and solutions of these materials were prepared in dust-free solvents, such as chloroform or hexane. 1 mg/ml solutions were used in order to spread monolayers of the amphiphiles on top of 12 MEGOHM purity water. The water troughs were stationed in a class 10 environment, in order to eliminate dust contamination on the water surface. A computer modified and controlled LAUDA film balance was used in all of the work.

A special design step dipper was used, which allowed 0.05 micron steps to be taken during the dipping process. The substrate was HMDS-treated, Si wafers (50 mm diameter). Up to 400 integrated devices had been previously implanted into the wafers, and these devices were used in the testing of film quality, reproduction, and device yield. The devices were prepared in an associated electrical engineering clean facility.

RESULTS AND DISCUSSION

The first studies dealt with the concept of a "processing window." A processing window is a narrow region of parameter space, wherein acceptable quality films are produced. The typical parameters in two dimensions might be collapse pressure, and temperature (Figure 1). As one can see from this figure, the region of acceptable quality is quite narrow, and requires careful measurement in order to determine this region accurately [1].

After an acceptable processing window is established, the next step is a careful study of the effect of dipping speed, and substrate roughness on film quality. This work is currently underway for all of the above materials. Initial results indicate that there is a new processing window available, associated with the dipping process alone.

Final film quality is affected by the way in which the film is handled and exposed to various environments after fabrication. Short exposures to non-clean conditions can have catastrophic effects on such important concepts as film stability and lifetime (Figure 2). This figure shows how a single dust particle, adsorbed onto the surface of a typical film (30 nm thick), can "eat" its way into the film, producing large holes [2].

The 168DA material was fabricated as the gate dielectric in as many as 400 FET's on a single wafer. The gates had varying gate widths and lengths in order to test the typical gate I-V characteristics of the FET. The new device, a MLSFET (Metal-Langmuir Film-Semiconductor-field-effect-transistor), produced I-V characteristics much in line with conventional silicon oxide dielectrics (Figure 3). These MLSFET's can, conceptually, serve as the basis of a number of novel new devices as well, such as chemical, pressure, and temperature sensors [3].

For the first time, an array of integrated field-effect transistors using Langmuir-Blodgett (LB) films as gate insulators have been fabricated and studied. The sources and drains of the FET's are prefabricated on (100) p-type silicon using conventional semiconductor processing.

Devices from early fabrication runs exhibited an initial positive drift of the drain current with time. The behavior of the drain current as a function of temperature suggested that the drift was due to mobile

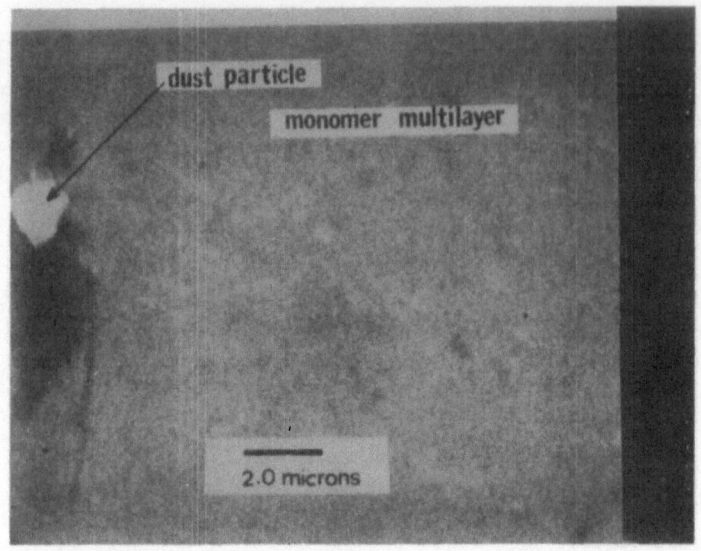

Figure 2 Scanning electron micrograph of dust particle on monomer multilayer (30 nm thick) and resultant damage.

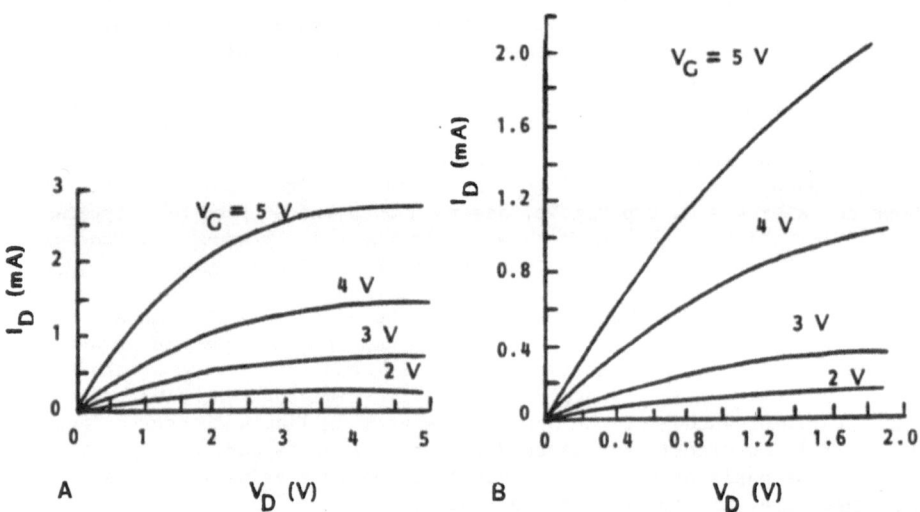

Figure 3. Curve tracer of various gate geometries of a typical MLSFET.
(A) The drain I-V characteristic of a typical IGFET obtained
from a curve tracer.
(B) I-V curves of the same device in the nonsaturation region.

charges in the film. The use of higher purity water during the LB deposition in later runs eliminated the drift.

By varying the polymerization conditions of the LB film, both X and Y-type LB films were produced. Capacitance-voltage measurements of the effective insulator charge confirms the existance of polarization in the X-type films.

Effective charge densities for devices with an X-type insulator were found to be on the order of -10^{11} q/cm^2. This is attributed to the polarization of the X-type film. The polarization was found to be 3.85×10^{-7} Coul/cm^2. The molecular dipole moment is calculated to be 2.98×10^{-27} Coul-cm. For normal, Y-type, insulators, the effective charge was on the order of 10^{11} q/cm^2. This is within an order of magnitude of the values reported for conventional metal-oxide-semiconductor devices. This charge appears to originate in the interface due to the presence of a thin native SiO$_2$ layer.

In measuring the current-voltage characteristics of the enhancement mode n-channel FET's, the surface mobility of the electrons was found to be independent of the effective insulator charge. The surface mobility at effective fields of 5×10^7 V/m was typically between 550 and 600 cm^2/V-sec compared to the 400 to 450 cm^2/V-sec values reported for Si-SiO$_2$ devices.

CONCLUSIONS

Workers in the PML have demonstrated an ability to control all important parameters needed to produce integrated Langmuir devices (ILD). There can now be no question as to the adaptability of organic films to electronics use. In the future, additional ILD's will be produced, if and when such devices can be prepared with high reliability, yield and operating lifetimes.

REFERENCES

1. M.B. Biddle, S.E. Rickert, J.B. Lando, "Constructing a Processing Window for a Langmuir-Blodgett Film," accepted by Thin Solid Films, (1985).

2. J.J. Spaulding, M.S. Thesis, "Electron-Beam Lithography of Ultrathin Langmuir-Blodgett Resists," Case Western Reserve University, August, 1985.

3. A.S. Dewa, C.D. Fung, E.P. DiPoto, S.E. Rickert, "The Study of Metal-Insulator-Semiconductor Structures with Langmuir-Blodgett Insulators," accepted by Thin Solid Films, (1985).

HIGH-RESOLUTION PHOTOREFRACTIVE POLYMERS FOR INTEGRATED OPTICS

W. Driemeir and A. Brockmekyer

Universität Osnabrück
FB Physik, PO Box 4469
D-4500 Osnabrück, F.R. Germany

INTRODUCTION

Some transparent polymers exhibit photorefractive behaviour: Their re-fractive-index can be changed by illumination with light. Hence these photo-refractive polymers are well suited for easy fabrication of passive compo-nents for Integrated Optics, if they fulfill some requirements:

1. high resolution   :   better than 1000 lines/mm
2. sufficient refractive-index change :   more than 10^{-2}
3. stability of the index patterns   :   several years   .

Waveguiding thin films of such organic polymers may be prepared by so-lution deposition or by spin-coating techniques. Subsequently refractive-index patterns forming optical waveguide circuits and waveguide gratings can be recorded optically by conventional mask printing, by computer controlled laser beam writing or by holographical techniques.

Recently a new photorefractive polymeric material was introduced by Franke/1/. This material consists of poly-(methyl methacrylate) ("PMMA") highly doped with the u.v.-photoinitiator 2,2-dimethoxy-1,2-diphenyl-ethanon ("DMDPE" or "benzildimethylketal"). Large index changes up to 0.05 were mea-sured for illuminated stripes of 1-2 mm width. Our holographical investiga-tions of this material, however, showed poor stability of high-resolution index patterns. Phase gratings with a period of 1 µm faded away in some days at room temperature or even in 5 min at 100°C.

Some years ago Tomlinson et al. /2,3/ introduced a high-resolution photolocking material suitable for applications in Integrated Optics. The polymeric matrix consisted of an unsaturated polymethacrylic ester. This polymer was doped with naphthalene-thiol-(2). Our investigations of this material, however, showed a clear tendency for crystallization of the dopant even during preparation of thin polymeric films. The resulting optical wave-guides often showed very high losses due to scattering. In addition this material is not very sensitive to near u.v.-radiation sources with a maximum intensity at 350-370 nm.

To realize a new high-resolution optical recording material for Inte-grated Optics we consequently doped the reactive polymer of Tomlinson et al.

with the photoinitiator DMDPE used by Franke. Waveguiding thin films of this material were investigated holographically and by mode spectra analysis.

PREPARATIONS

Synthesis of Reactive Polymer

A polymethacrylic ester with unsaturated side chains is synthesized in two steps (see Fig. 1) /3,4/:

1. Copolymerization of methyl methacrylate (MMA) and glycidyl methacrylate (GMA)
2. Esterification between the epoxy-groups of the copolymer and monobasic unsaturated caboxylic acids like cinnamic acid and mono-ethyl fumarate.

recipe:	polymer type A	polymer type B
MMA	1.00 mol	1.00 mol
GMA	1.00 mol	1.00 mol
mono-ethyl fumarate	1.00 mol	0.75 mol
cinnamic acid	---	0.25 mol

Preparation of Photorefractive Planar Optical Waveguides

Planar optical waveguides are prepared by solution deposition of thin polymeric films.

recipe:	dry unsaturated polymethacrylic ester	0.400 g
	dopant DMDPE	0.400 g
	methyl methacrylate	50.00 ml

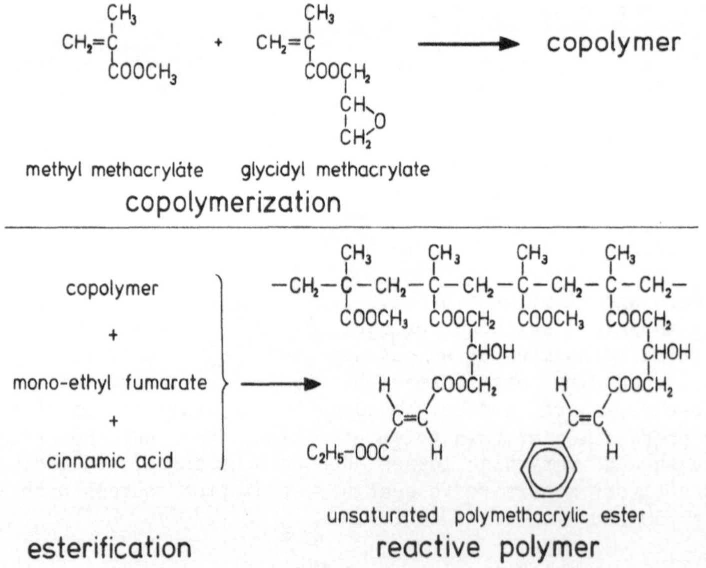

Fig. 1 Synthesis of an unsaturated polymethacrylic ester.

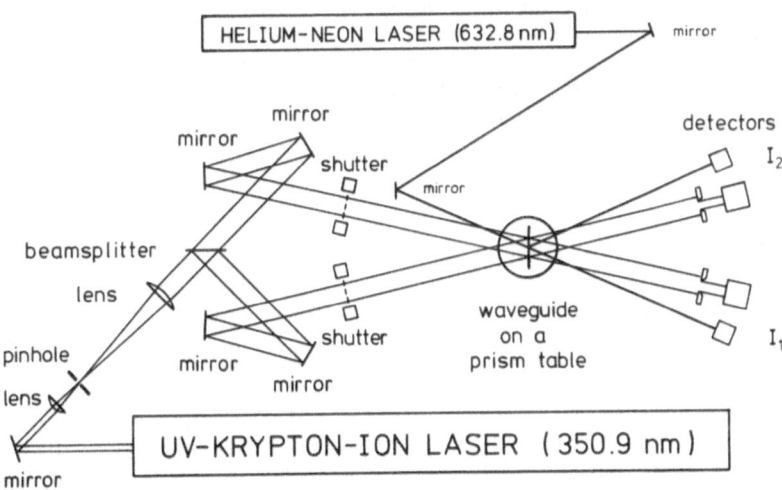

Fig. 2 Holographical set-up.

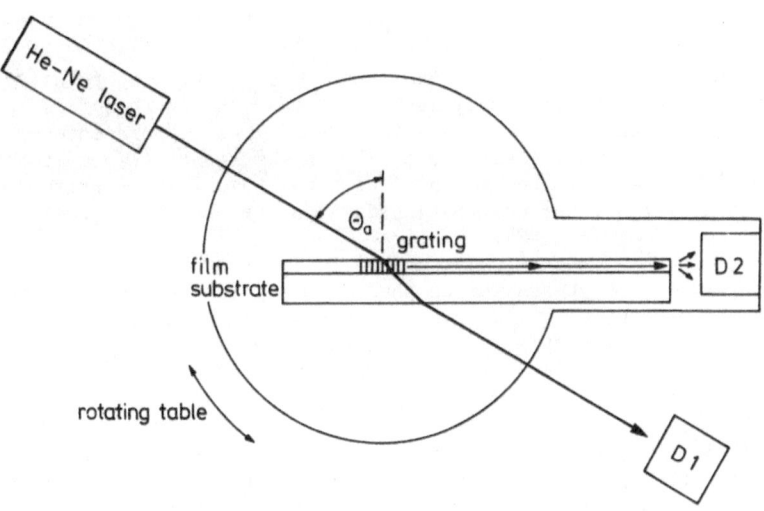

Fig. 3 Mode spectra analysis by the grating coupler method.

A quartz glass slide is coated with 1.0 ml of this solution. Slow controlled evaporation of the solvent in a desiccator yields a dry thin film with a smooth surface. The thickness is about 5 μm. Because the index of this film is higher than the index of the slide, visible light can be guided inside the film by total internal reflection. This sandwich structure represents a planar optical waveguide. The refractive-index of the organic film can be raised locally by illumination with u.v.-light under formation of the desired index pattern.

INVESTIGATIONS OF THE PHOTOREFRACTIVE PROPERTIES

Holographical Investigations

Holographical methods are very efficient to investigate the properties of photorefractive materials. The holographical set-up is shown in Fig. 2. The coherent u.v.-radiation source is a Krypton ion laser (350.9 nm). Optical phase gratings are recorded inside the photorefractive thin films by exposure to the interference field of two expanded laser beams. The formation of the phase gratings is monitored continuously with a Helium Neon laser. The time-dependent diffraction efficiency η of the grating can be defined as

$$(1) \qquad \eta = I_2 / (I_1 + I_2)$$

I_2 is the intensity of the diffracted beam, I_1 is the intensity of the transmitted beam. From the measured diffraction efficiency the light-induced amplitude of the index modulation n_1 can be determined according to KOGELNIK's coupled wave theory /5/:

$$(2) \qquad \eta = \sin^2 ((\pi n_1 d)/(\lambda \cos \Theta_f)) \qquad ,$$

where d is the thickness of the film, λ the free space wavelength, and Θ_f the Bragg angle inside the film. The peak-to-peak index change of a sinusoidal index grating is given by $\Delta n = 2n_1$.

Mode Spectra Analysis

Thickness d and refractive-index n_f of a light-guiding thin film can be determined by mode spectra analysis. Light-guiding in thin films is only possible in certain electromagnetic modes. These "guided modes" can be excited with an external laser beam. For this purpose it is common to press a high-index prism onto the surface of the thin film ("prism coupler"). Unfortunately very often deformation and damage of the organic thin film occur.

However, if a high-resolution phase grating is written holographically into the photorefractive film, the grating itself can be used as coupling structure ("grating coupler"). This method needs no contact between the organic thin film and a bulky prism. Hence deformation and damage of the film are avoided /6/.

Excitation of guided modes happens only, if the external laser beam strikes the grating region at certain "synchronous" angles. Mode spectra analysis means the determination of these discrete angles. For that purpose the optical waveguide is mounted on a motorized rotating table (see Fig. 3). The excitation of the guided modes is detected with a photodiode at the end of the waveguide. From the related synchronous angles refractive-index and thickness of the thin film can be calculated numerically /6/.

EXPERIMENTAL RESULTS

Formation of High-Resolution Phase Gratings

A typical holographical growth curve of a photorefractive thin film is shown in Fig. 4a. The formation of the phase grating with a period of 0.804 µm starts immediately at the beginning of the illumination. An initial delay time cannot be observed. After a continuous increase the diffraction efficiency η reaches saturation.

Mode spectra analysis yields refractive-index and thickness of the organic thin film. The light-induced refractive-index change Δn in the grating region can then be calculated according to KOGELNIK's formula (2) (see Fig. 4b). After 8 minutes of illumination a maximum refractive-index change of $2.1*10^{-2}$ is achieved. But this value deteriorates in some days due to diffusion and relaxation processes.

To improve the stability of the refractive-index patterns it is convenient to apply only short illumination times up to 60 s corresponding to an exposure of about 1 J/cm². The initial diffraction efficiency immediately after illumination may be low, but this behaviour corresponds to the formation of a latent image. Simple annealing at 100°C for some hours increases the refractive-index changes remarkably (see Fig. 5). Applying this thermal developing process phase gratings with a stable refractive-index change up to $1.8*10^{-2}$ has been produced.

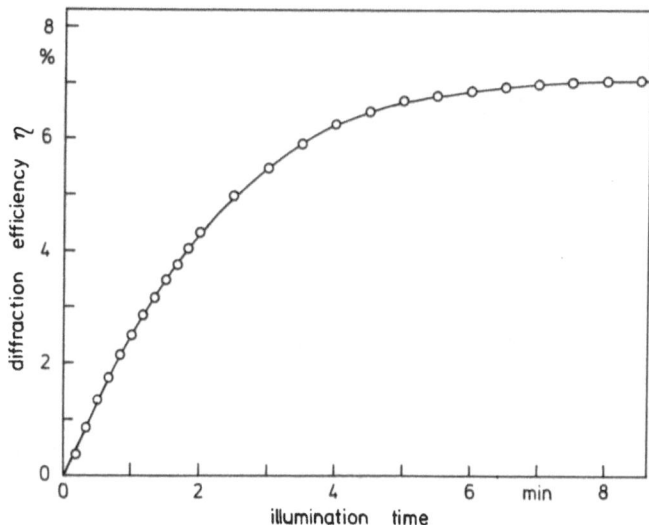

Fig. 4a   Holographical recording of an optical phase grating :
Diffraction efficiency vs. illumination time.

Film parameters: 50 wt% polymer type B / 50 wt% DMDPE, thickness d = 4,7 µm, index n_f = 1.543, grating period Λ = 0.804 µm, intensity $2I_0$ = 30 mW/cm².

247

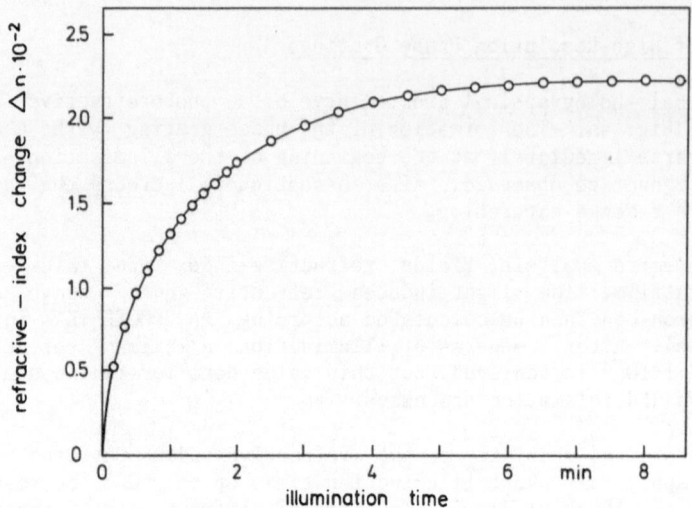

Fig. 4b Holographical recording of an optical phase grating :
Refractive-index change vs. illumination time.

Evaluation for the organic thin film of Fig. 5a.

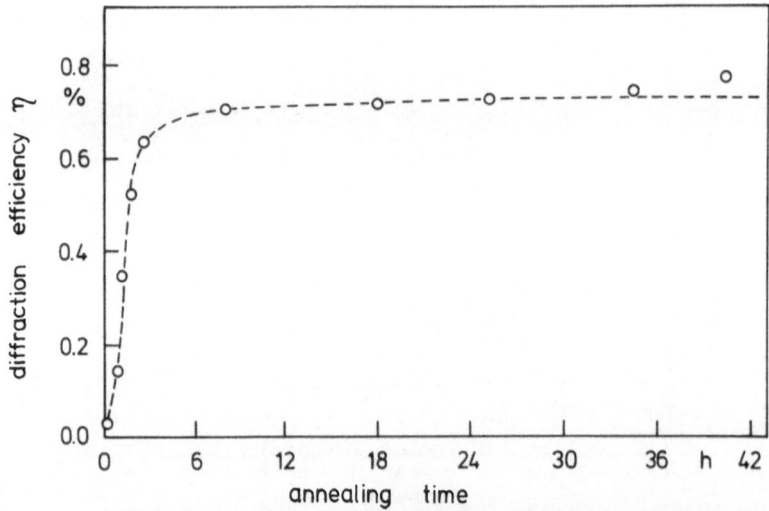

Fig. 5 Thermal development of a holographical phase grating:
Diffraction efficiency vs. annealing time.

Film parameters: 50 wt% polymer type B/ 50 wt% DMDPE,
d = 1.03 µm after annealing at 100°C , Λ = 0.803 µm,
$2I_0$ = 30 mW/cm^2 for 60s.

Stability of the Refractive-Index Patterns

A typical disadvantage of most photorefractive polymers is the low stability of the recorded phase patterns. Our reactive polymeric system, however, exhibits remarkable stability of the recorded phase patterns after the thermal development. To examine the thermal stability a high-resolution phase grating is submitted to a thermal stress test (see Fig. 6). The phase grating has been annealed in addition for 12 days at increasing temperatures starting at 120°C. The temperature is raised every day about 10°C. The total breakdown of this grating occurs not before 230°C.

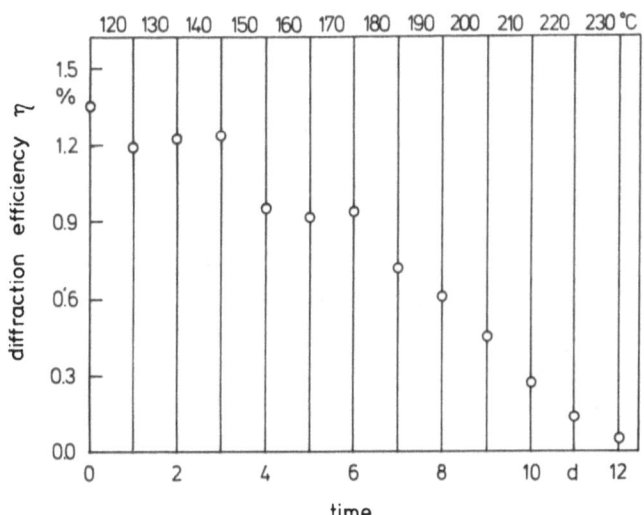

Fig. 6   Thermal stress test of a holographical phase grating. Film parameters: 50 wt% polymer type B / 50 wt% DMDPE, $d = 5$ μm, $\Lambda = 0.829$ μm, $2I_0 = 30$ mW/cm^2 for 60s.

DISCUSSION OF THE MECHANISM OF REFRACTIVE-INDEX CHANGE

The sensibility of the organic thin films in the near u.v.-region is caused only by the high content of the photoinitiator DMDPE. Under u.v.-irradiation these molecules undergo a fragmentation reaction of Norrish-type I forming free radicals /7/. These radicals are fixed to the polymer by radical addition to the double bonds of the unsaturated side chains. It results a light-induced fixing of the high-index dopant to the low-index polymer ("photofixing") (see Fig. 7). If there is still a small content of the reactive solvent MMA in the organic thin films, an additional light-induced crosslinking hardens the polymeric matrix ("photocrosslinking"). These two reactions seem to be the reason for the remarkable stability of the refractive-index patterns.

If the organic thin films are illuminated shortly, the resulting index patterns are nearly latent because there is still a high content of unreacted photoinitiator in the unilluminated regions. The photoinitiator DMDPE, however, is moderately volatile at 100°C. For that reason annealing at 100°C for some hours leads to evaporation of unfixed photoinitiator. It remains only the low-index polymer in the unilluminated regions. By this thermal development the index differences between illuminated and unilluminated regions are enhanced.

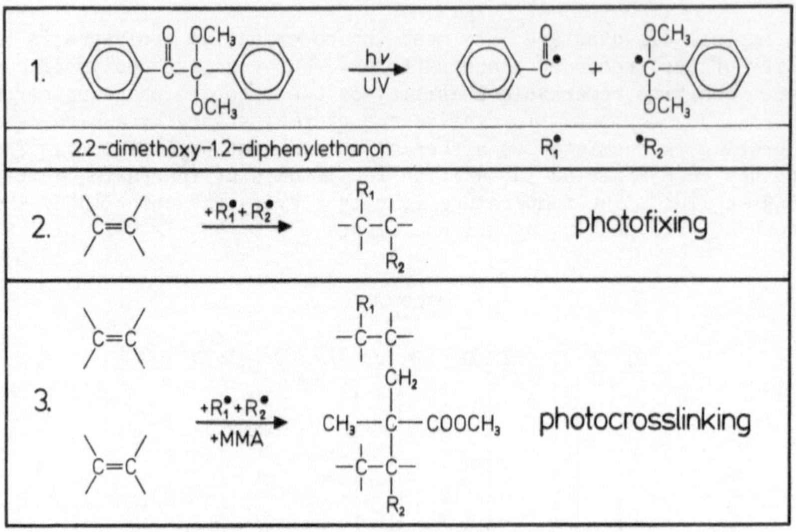

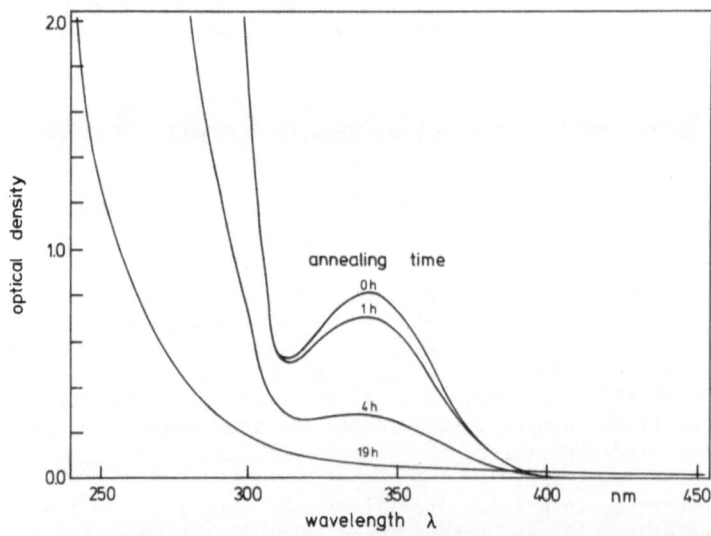

Fig. 7　Reaction mechanism :
1. Light-induced generation of free radicals.
2. Photofixing of high-index radicals to the low-index polymer.
3. Photocrosslinking of the polymeric matrix.

Fig. 8　Absorption spectra of an unilluminated film during annealing at 100°C.

Film parameters: 50 wt% polymer type A / 50 wt% DMDPE, d = 9.52 µm, n_f = 1.5405 before annealing; d = 4.29 µm, n_f = 1.5081 after annealing for 19h.

The evaporation of the photoinitiator during annealing can be demonstrated by monitoring the changes of the absorption spectrum of an unilluminated thin film (see Fig. 8). According to the evaporation of DMDPE the broad peak near 340 nm vanishes. There remains only the absorption spectrum of the undoped polymeric matrix.

CONCLUSIONS

The above described photorefractive polymer shows

1. high resolution   : better than 1200 lines/mm
2. refractive-index change   : $\Delta n = 1.8*10^{-2}$
3. remarkable stability of the index patterns according to the thermal stress test.

Consequently this material is well suited for easy fabrication of passive components for Integrated Optics. Waveguiding structures can be recorded optically by conventional photolithography, by writing or by scanning with a focussed u.v.-laserbeam or by holographical methods. No wet processing is needed after illumination. So this photorefractive polymer shows good properties as optical recording material to realize refractive-index patterns with high resolution.

Acknowledgements

This work is part of a cooperation between the groups for Applied Physics and Physical Chemistry at the University Osnabrück. We wish to thank E.Krätzig and M.D.Lechner for their help and support. Our further thanks are due to W.Wunderlich and H.J.Auer of Fa.Röhm, Darmstadt for valuable discussions. The financial support of the Stiftung Volkswagenwerk is gratefully acknowlegded.

REFERENCES

1. H. Franke, Optical Recording of Refractive-Index Patterns in Doped Poly-(methyl methacrylate) Films, Appl. Opt. 23: 2729 (1984)
2. W. J. Tomlinson, H. P. Weber, C. A. Pryde, and E. A. Chandross, Optical Directional Couplers and Grating Couplers Using a New High-Resolution Photolocking Material, Appl. Phys. Lett. 26: 303 (1975)
3. W. J. Tomlinson and E. A. Chandross, Organic Photochemical Refractive-Index Image Recording Systems, Adv. Photochem. 12: 201 (1980)
4. E. H. S. Nal and G. Smets, Photocrosslinking of Unsaturated Polymethacrylic Esters, Polym. Photochem. 5: 93 (1984)
5. H. Kogelnik, Coupled Wave Theory for Thick Hologram Gratings, Bell Syst. Tech. J. 48: 2909 (1969)
6. W. Driemeier and A. Brockmeyer, Grating Coupler Method for the Characterization of Photorefractive Thin Films, submitted to Opt. Comm.
7. R. Kirchmayr, G. Berner, and G.Rist, Photoinitiatoren für die UV-Härtung von Lacken, farbe+lack 86: 224 (1980)

SOLAR ENERGY CONVERSION BY MEANS OF PARAMETRIC MECHNISMS

Herbert Wetzel and Helmut Tributsch

Hahn-Meitner-Institut Berlin, Bereich Strahlenchemie
1000 Berlin 39, Federal Republic of Germany

The theories of energy conversion by photovoltaic, photoelectro-chemical, photochemical and solar thermal processes involving light induced charge carrier separation or photon excitation, respectively, are well established. Although parametric mechanisms play a crucial role in many branches of physics, mostly as undesired, energy consuming side effects, up to now no attempt has been made to investigate the applica-bility of these processes for solar energy conversion.

The principle of parametric excitation is based on the periodic variation of an energy definining parameter of a system capable of oscillations.[1] By periodic illumination of an "active" device solar energy can be pumped into a system. Among the various candidates for parametric excitation oscillating electric circuits may be regarded as promising ones, since the capacity or the inductivity might easily be constructed as photosensitive active parts of the circuit. In this case parametric excitation is performed by chopped illumination of either the photosensitive capacity or inductivity.

This is described by taking into account the time dependence of the capacity or the inductivity in the differential equation for an oscillating electric circuit. For example:

$$L\ddot{q} + R\dot{q} + \frac{1}{C(t)} q = 0$$

L, R, and q have the usual meanings

$$C(t) = \begin{cases} C_0 - C & \text{for } 0 < t < \frac{T}{4}, \frac{T}{2} < t < \frac{3T}{4} \\ C_0 & \text{for } \frac{T}{4} < t < \frac{T}{2}, \frac{3T}{4} < t < T \end{cases}$$

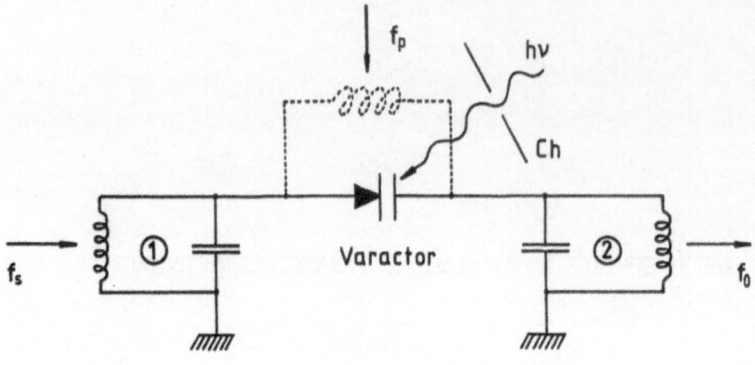

Figure 1

with $T = \dfrac{2}{\omega_0}$ and $\omega_0 = \dfrac{1}{LC}^{1/2}$ (eigenfrequency of the circuit)

If $\dfrac{\Delta C}{C} < 2\ R\left(\dfrac{C}{L}\right)^{1/2}$

instable periodic oscillations with increasing amplitude will arise.

It is evident that the efficiency of parametric energy conversion depends on the degree of parametric change upon illumination. Thus materials exhibiting highly nonlinear photoresponse are required.

In order to demonstrate that, in principle, it is possible to excite light induced parametric mechanism we introduce a light induced parametric frequency modulator (see figure).

Variation of the capacitance at frequency f_p by illumination causes a signal which can be modulated by a small input signal f_s. At the output signals at f_p $n f_s$, $n = 1, 2 \ldots$ appear. We performed our experiment with $f_p = 100$ kHz and $f_s = 20$ kHz. The oscillating circuits have been tuned to 20 kHz and 120 kHz, respectively. At the output a 120 kHz signal was available, intensity of which was dependent on the degree of capacitance variation and on the degree of non-linearity.

From this result we conclude that not only photoparametric modulation but also photoparametric amplification, which means that ($I f_0$ (I = intensity) higher than $I f_s$) might be achieved by optimization of the varactor characteristic, i.e. by a highly non-linear capacitance/voltage characteristic and high variation of the capacitance upon illumination.

REFERENCES

1. H. Wetzel and H. Tributsch, Exploration of Parametric Mechanisms of Photon Energy Conversion, Solar Energy, 37 (1986) 65

AUTHOR INDEX

SUBJECT INDEX

Absorption spectrum, 149
Acceptor, 117,119,120,128,130,153-
    155,157,160,163,174,204
Accumulation point, 2
Activation energy, 146
Affine map, 8
Amorphous silica, 185
Amorphous solid, 57
Annihilation rate, 12,25
Anthracene, 63,68,75,77-79,84
Attractor, 1,2,5,6,8

Band edge, 204,207,226
Band tail, 210
Band tail states, 207
Bare exciton, 112
Birefringence, 67
Bond coupling, 120

Cellular automata, 1,2,9
Change transfer, 123
Chaos game, 2,3
Charge transfer, 122,145,146,
    148,151
Chemisorption, 136,142,172
Collage theorem, 6
Conducting polymers, 151
Coupling, 123
Crystal damage, 166
Crystalline silicon, 227

Dangling bond, 198,202
Data compression, 5
Debye-Waller factor, 187
Defect accumulation, 166
Dendritic crystal, 2
Density of states, 200,210
Dielectric loss, 134
Dielectric media, 126
Diffusion constant, 227,228
Disordered material, 153
Domain, 64

Domain size, 14,27
Donor, 117,119,120,123,128,130,
    154,155,157,160,163,204

Electrical conductivity, 145
Electrochromism, 149,150
Electron transfer, 117,130
Electron-hole pairs, 226,229,233
Electron-hole recombination, 17
Enthalpy, 140
Excimer, 83,86,87,90,96,97,179-
    182,187
Exciton annihilation, 21
Exciton fusion, 63
Exciton, 12,21,27,63,80,106,107,
    109-113,115,122,123

Fluorescence, 16,22,23,55,63-66,
    69,71-73,75-80,83,86-92,94,
    98-102,173,174,179,185,186,
    194
Fluctuation-dissipation relation,
    112
Fractal, 1,2,11-14,17,30,34,63,
    64,80,97,153,154,157,163,
    171,173,176,177
Fractal domain, 22
Free energy, 140,141
Free radical, 250
Frenkel exciton, 105,106
Fusion rate, 13

GaAlAs laser, 45
Gated PHB, 47-50
Grain boundary, 11

Hole burning, 41,45,47,50,55,168,
    169,185,193
Hole hopping, 148
Hole transfer, 123
Hole trapping, 138